MICROCONTROLLERS

Fundamentals and Applications with PIC

MICROCONTROLLERS

Fundamentals and Applications with PIC

Fernando E. Valdes-Perez
Ramon Pallas-Areny

CRC Press
Taylor & Francis Group
Boca Raton London New York

CRC Press is an imprint of the
Taylor & Francis Group, an **informa** business

CRC Press
Taylor & Francis Group
6000 Broken Sound Parkway NW, Suite 300
Boca Raton, FL 33487-2742

© 2009 by Taylor & Francis Group, LLC
CRC Press is an imprint of Taylor & Francis Group, an Informa business

---

Library of Congress Cataloging-in-Publication Data

Valdés Pérez, Fernando E.
  Microcontrollers : fundamentals and applications with PIC / authors, Fernando E. Valdes-Perez and Ramon Pallas-Areny.
    p. cm.
  Includes bibliographical references and index.
  ISBN 978-1-4200-7767-4 (alk. paper)
  1. Programmable controllers. 2. Microcontrollers. I. Pallàs-Areny, Ramón. II. Title.

TJ223.P76V346 2009
629.8'9--dc22   2008044213

Visit the Taylor & Francis Web site at
http://www.taylorandfrancis.com

and the CRC Press Web site at
http://www.crcpress.com

Contents

Preface

Microcontrollers are present in most products of daily use. Teaching microcontrollers is difficult because of the wide variety of models, which are based on different structures, as well as the large number of their possible applications. Despite this diversity, it is possible to find common elements in the architecture of most microcontrollers. This book exploits these common elements to describe the fundamentals of microcontroller design and programming.

This book aims to help the reader learn the architecture and programming of generic microcontrollers using the programmable integrated circuit (PIC) family from Microchip as examples. The documentation provided by the manufacturers of these devices is extensive and it can become overwhelming. The topics in this book have been chosen in such a way to ensure their continuity, focusing on the clear and accurate explanation of these concepts. We have included figures that add value to the book, and we have avoided pictures or other graphic material that, while increasing the number of pages, do not add any substantial information. Moreover, these pictorial materials can be easily found on the manufacturers' Web sites. To help the learner, the first time a new term is introduced, it is in italics.

Each topic is treated using a reader-centered, top-to-bottom approach. First, we expose and describe the issues that are common to any microcontroller. Afterward, these topics are studied in detail for PIC microcontrollers. The book has a large number of examples that are taken from real-life applications, thus reinforcing the concepts and relating them to industry.

This book is structured in nine chapters. Chapter 1 describes the structure and resources of a generic microcontroller. Chapter 2 describes PIC microcontrollers with a special focus on medium-end devices. Chapter 3 explains the memory organization and structure of microcontrollers in general, focusing again on medium-end PICs. Chapter 4 describes the assembler language used for programming medium-end PIC microcontrollers. Assembler language is the best option for relatively simple applications in which the microcontroller needs to execute small tasks using simple algorithms. The use of assembler language minimizes the amount of memory needed, thus allowing the selection of a smaller microcontroller. When faced with complex algorithms, the best programming option becomes high-level programming language. This requires the use of compilers that are not always free.

Chapters 5, 6, 7, and 8 describe how microcontrollers can acquire, process, and generate digital signals. These chapters explain available techniques to deal with parallel input or output, peripherals, resources for real-time use, interrupts, and the specific characteristics of serial data interfaces in PIC microcontrollers. Chapter 9 describes the acquisition and generation of analog signals either using resources inside the chip or by connecting peripheral circuits.

The appendix contains a list of acronyms used. The final pages contain bibliographical references for those readers who may desire to deepen their knowledge of these topics.

This book is aimed toward electronics students and professionals, but it will also be useful for those readers interested in learning more about PIC microcontrollers and how to use them efficiently.

Fernando E. Valdés Pérez

Ramon Pallàs-Areny

The Authors

Fernando Eudaldo Valdés Pérez received his BS and MS degrees in electrical engineering from the Universidad de Oriente in Cuba in 1977 and 2001, respectively. He is an associate professor at the Center of Neuroscience Studies and Image and Signal Processing at the Universidad de Oriente. He has broad teaching experience, mostly focused on architecture programming of microprocessors, microcontrollers, and personal computers, as well as the statistical treatment of signals for biomedical applications. He is the main author of the textbook *Fundamentos Técnicos de Computación* (Fundamentals of Computing; ISPJAE, La Habana, 1986). His current research is focused on the acquisition and processing of cardiovascular signals. He has also worked on the design of hemodialysis monitoring systems.

Ramon Pallàs-Areny received the Ingeniero Industrial and Doctor Ingeniero Industrial degrees from the Universitat Politècnica de Catalunya (UPC), Barcelona, Spain, in 1975 and 1982, respectively. He is a professor of electronics engineering at the same university, and teaches courses in electronics and medical instrumentation. In 1989 and 1990 he was a visiting Fulbright Scholar, and in 1997 and 1998 he was an Honorary Fellow at the University of Wisconsin, Madison. His research includes instrumentation methods and sensors based on electrical impedance measurements, autonomous sensors, sensor interfaces, noninvasive physiological measurements and electromagnetic compatibility in electronic systems. He is the author of six books, the leading author of five books, and coauthor of two books on instrumentation in Spanish and Catalan. He is also coauthor, with John G. Webster, of *Sensors and Signal Conditioning*, 2nd edition (New York, Wiley, 2001), and *Analog Signal Processing* (New York, Wiley, 1999); and with Ferran Reverter on *Direct Sensor-to-Microcontroller Interface Circuits* (Barcelona, Marcombo, 2005). Dr. Pallàs-Areny was a recipient, with John G. Webster, of the 1991 Andrew R. Chi Prize Paper Award from the IEEE/Instrumentation and Measurement Society. In 2000 he received the Award for Quality in Teaching granted by the Board of Trustees of UPC, and in 2002 the Narcís Monturiol Medal from the Autonomous Government of Catalonia.

1

Introduction to Microcontrollers

This chapter studies the structure and resources found in typical micro-controllers. It starts by introducing the concept of a microcontroller and exploring the differences between microcontrollers and microprocessors. It continues with the description of the resources that are available in micro-controllers, focusing again on how they differ from the resources available in microprocessors. The chapter then describes the von Neumann and Harvard architectures as well as how the reduced instruction set computer (RISC) and complex instruction set computer (CISC) architectures differ in their instruction sets. It finishes by describing the most common microcontrollers and listing their manufacturers.

1.1 Microprocessors and Microcontrollers: Characterization

Figure 1.1 shows the block diagram of a generic microcomputer. It consists of three fundamental blocks: central processing unit (CPU), memory, and input/output (I/O) system. These blocks are interconnected by groups of electrical lines called *buses*. The buses that transport memory or I/O addresses are called *address buses*; the buses that transport data or instructions are called *data buses*; and the buses that transport control signals are called *control buses*.

The CPU is the brain of the microcomputer, being under control of the program stored in memory. The tasks of the CPU are to fetch the instructions stored in memory, interpret those instructions, and execute them. The CPU also includes the circuitry necessary to perform arithmetic and logic operations with binary data. This special circuitry is called the arithmetic and logic unit (ALU).

In a microcomputer, the CPU is its *microprocessor*, which is the integrated circuit that carries out the operations described above. A microcontroller can be considered as a microcomputer built on a single integrated circuit or *chip*. Historically, microcontrollers appeared after microprocessors and followed independent paths. Microprocessors are mainly found in personal computers and workstations, as these require strong computational power, and the ability to manage large sets of data and instructions at a high speed. A very important parameter for microprocessors is the size of

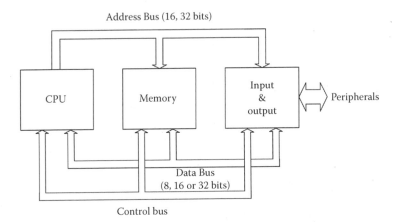

FIGURE 1.1
Generic block diagram of a microcomputer. Here, the CPU is the microprocessor.

their internal registers (8, 16, 32, or 64 bits), as this determines the number of bits that can be processed simultaneously.

On the other hand, microcontrollers are used in a large variety of applications. They can be found in the automotive industry, communication systems, electronic instrumentation, hospital equipment, industrial equipment and applications, household appliances, toys, and so forth. Microcontrollers have been designed to be used in applications in which they have to carry out a small number of tasks at the lowest possible economic cost. They do this by executing a program permanently stored in their memory, whereas the input/output ports of the microcontroller are used to interact with the outside world. Therefore, the microcontroller becomes part of the application; it is a controller embedded in the system. Complex applications can use several microcontrollers, each one of them focusing on a small group of tasks.

The following generic requirements are important for microcontrollers and designs using microcontrollers:

1. Input/output resources. As opposed to microprocessors in which the emphasis is on computational power, microcontrollers put their emphasis on their input/output resources, such as the ability to handle interrupts, analog signals, number of different input and output lines, and so forth.

2. Optimization of space. It is important to use the smallest possible footprint at a reasonable cost. Given that the number of pins in a chip depends on its packaging, the footprint can be optimized by having one pin able to perform several different functions.

3. Using the most appropriate microcontroller for a given application. Microcontroller manufacturers have developed families of devices with the same instruction set but different hardware aspects, such as memory size, input/output devices, and so forth. This allows the designer to select the most appropriate device from a given family.

4. Protection against failure. It is critical for safety to guarantee that the microcontroller is executing the correct program. If for any reason the program goes astray, the situation has to be immediately corrected. Microcontrollers have a *watchdog timer* (WDT) to ensure that the program is being executed correctly. Watchdog timers do not exist in personal computers.

5. Low power consumption. Because batteries power many applications using microcontrollers, it is important to ensure the low power consumption of microcontrollers. Furthermore, the energy used when the microcontroller is not doing anything, for example, when it is waiting for an action from the user like a keyboard input, needs to be kept to a minimum. To do this, the microcontroller is set in sleeping state until it resumes the execution of the program.

6. Protection of programs against copies. The program stored in memory needs to be protected against unauthorized reading. To do this, the microcontrollers incorporate protection mechanisms against copying.

1.2 Components of a Microcontroller

Microcontrollers combine the fundamental resources available in a microcomputer such as the CPU, memory, and I/O resources in a single chip. Figure 1.2 shows the block diagram for a generic microcontroller.

Microcontrollers have an oscillator to generate the signal necessary to synchronize all internal operations. Although this can be a basic *RC* (resistance capacitor) oscillator, a quartz crystal (XTAL) is normally used due to its high frequency stability. The frequency of the oscillator has a direct influence on the speed at which program instructions are executed.

Similar to microcomputers, the CPU is the brain of the microcontroller. The CPU fetches the program instructions from their locations in memory one by one, interprets or decodes them, and executes them. The CPU also includes the ALU circuits for binary arithmetic and logic operations.

The microcontroller's CPU has different registers. Some of these registers are intended for general use, whereas others have a specific pur-

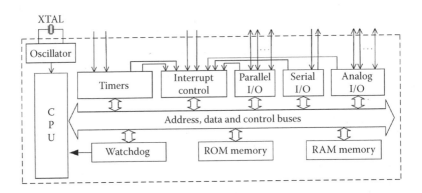

FIGURE 1.2
Basic block diagram of a microcontroller.

pose. Specific purpose registers include: instruction register, accumulator, status register, program counter, data address register, and stack pointer.

The *instruction register* (IR) stores the instruction that the CPU is executing. The programmer does not normally have access to the IR.

The *accumulator* (ACC) is a register associated with the arithmetic and logic operations that the ALU is carrying out. When executing any operation, one of the data needs to be in the ACC. The resulting value is also stored in the ACC. PIC microcontrollers do not have the ACC register. Instead, they have a working (W) register that is very similar to the ACC.

The *status register* (STATUS) contains the bits that show different characteristics related to the operations carried out by the ALU. These can be the sign of the resulting value (positive vs. negative), a flag to notify if the resulting value is zero, carry over, parity bits, and so forth.

The *program counter* (PC) is the CPU register where addresses of instructions are stored. Every time that the CPU looks for an instruction in the memory, the PC is increased, pointing to the following instruction. In an instant of time, the PC contains the address of the instruction that will be executed next. The control transfer instructions modify the value stored in the PC.

The *data address register* (DAR) stores data addresses from memory. This register plays a major role in indirect data addressing. Different types of microcontrollers use different specific names for the DAR. For example, PIC microcontrollers call this register the file select register (FSR).

The *stack pointer* (SP) stores data addresses in the stack. The stack and the SP register are studied in further detail in Chapter 4. PIC microcontrollers do not have an SP register.

The microcontroller memory stores both program instructions and data. Any microcontroller has two types of memory: random-access memory (RAM) and read-only memory (ROM). RAM can be read and written.

RAM is volatile memory, meaning that its data is lost when it is not powered. On the other hand, although ROM can only be read, it is non-volatile. The different types of technologies used for ROM such as EPROM (erasable programmable read-only memory), EEPROM (electrical erasable programmable read-only memory), OTP (one-time programmable), and FLASH are described in detail in Chapter 3. Both RAM and ROM are "random access" memories, meaning that the time to access specific data does not depend on its stored location. This is opposed to sequential access memories in which the time needed to access a specific memory cell depends on the location of the last accessed cell.

ROM is used to permanently store the program for the microcontroller, whereas RAM is used to temporarily store the data that will be manipulated by the program. An increasing number of microcontrollers use nonvolatile memory such as EEPROM to store some of the data that is changed only sporadically. The size of ROM is larger than the size of RAM for two main reasons: First, most applications require programs that manipulate a relatively small number of data. Second, RAM has a larger footprint compared to ROM, and therefore it is more expensive than ROM.

Being the vehicle to communicate with the outside world, the I/O resources are very important in microcontrollers. I/O resources consist of the serial and parallel ports, timers, and interruption managers. Some microcontrollers also incorporate analog input and output lines associated with analog-to-digital (A/D) and digital-to-analog (D/A) converters. The resources needed to ensure the regular operation of the microcontrollers such as the watchdog are also considered part of the I/O resources.

Parallel ports are normally structured in groups of up to eight lines of digital inputs and outputs. It is normally possible to manipulate each one of these lines individually. Serial ports can be of different technologies such as RS-232C (Recommended Standard 232, Revision C), I²C (inter-integrated circuit), USB (universal serial bus), and Ethernet. In general, a microcontroller will have the largest possible number of I/O resources for the number of available pins in its integrated circuit package. To increase the performance, one physical pin can be connected to several internal blocks, and therefore that pin may carry out different functions depending on how the microcontroller has been configured.

1.2.1 The Watchdog

The watchdog timer (WDT) is a resource that can be found in most microcontrollers. As shown in Figure 1.3, the WDT consists of an oscillator and a binary counter of N bits. Although the oscillator can be the same oscillator used by the microcontroller, it is preferable to use an independent oscillator. The output of the counter is connected to the *reset input* for the microcontroller. The counting process can never be stopped, although the program being executed can periodically reset the counter to its initial value.

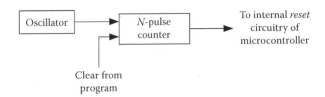

FIGURE 1.3
Basic watchdog diagram. Its output is connected to the internal circuitry to generate the reset signal.

Every pulse at the output of the oscillator becomes an input to the counter. When the counter reaches its maximum value, the output of the counter becomes active and gives a reset signal to the microcontroller. The goal of the designer is to avoid having the counter in the WDT reach its maximum value. Because, once started, the WDT cannot be stopped; the only way to avoid the reset signal is by setting the counter back to zero from the program that is being executed. This has to happen periodically and faster or the WDT counter will reach its maximum value. When the program is executed correctly, the WDT counter will never reach the maximum value. However, if the program becomes lost and stops executing the program, the WDT counter will reach its maximum value, will send the reset signal to the microcontroller, and the program will start executing from the beginning again. Therefore, the WDT is a critical element in a microcontroller, as it guarantees that the program will be executed continuously.

1.2.2 Reset Signal

Reset is an action that initializes microprocessors and microcontrollers. This initialization happens when a specific signal (called the *reset signal*) is applied to a specific pin (called the *reset pin*). The reset signal sets the program counter (PC) to a predetermined value, for example, PC = 0, making the microprocessor or microcontroller start executing the program commands from that specific memory address.

In a microcomputer, the reset signal can be applied manually (for example, when pressing a reset push button) or when the microcomputer boots up (*power-on reset*). Figure 1.4 shows the schematic of a circuit used to generate the reset signal either manually or through power-on. In the figure, V_{RESET} is the voltage applied to the reset pin, and V_{TH} is the threshold voltage for the pin. If $V_{RESET} < V_{TH}$, the device understands it as a logic value of 0 (RESET = 0). If $V_{RESET} > V_{TH}$, the device understands it as a logic value of 1 (RESET = 1). As shown in the schematic, the reset action occurs when RESET = 0.

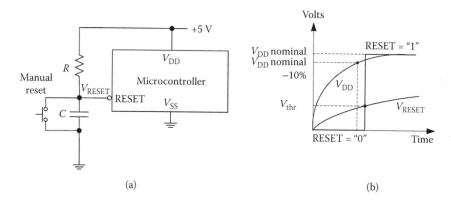

FIGURE 1.4

Manual reset and power-on reset. (a) Typical reset circuit in a microcontroller. (b) Time evolution of voltages involved in reset signal.

The resistance (R) and the capacitor (C) make up a simple RC circuit with a time constant $\tau = RC$. In power-on, the voltage supply charges C through R. If τ is large enough, V_{RESET} is lower than V_{TH} during the time it takes for the voltage supply to become stable for the microcontroller to work correctly.

In addition to these two voluntary reset actions, the microcontroller may be unwillingly reset, for example, due to problems with its power supply (*power-glitch reset, brown-out reset*) or due to an action from the WDT. Power-glitch reset occurs when the applied voltage momentarily falls below a certain value, so the capacitor discharges to the point that $V_{RESET} < V_{TH}$. The reset originated by the WDT occurs when the WDT has not been refreshed and the output of the counter becomes active. This normally happens when the microcontroller has stopped executing the correct program in memory. It is very important for the microcontroller to generate a reset signal in this case to guarantee that the microcontroller will start executing instructions from a known memory address instead of reaching an unknown memory location that could damage the system. Some microcontrollers utilize specific bits in a register to signal that a reset action has taken place. This allows us to further investigate the reasons as to why the reset took place in order to take corrective actions.

1.2.3 Low Consumption

Because batteries power most applications using microcontrollers, power consumption has become a critical parameter. Power consumption in an integrated circuit depends on three factors: the technology used in the chip, the frequency of its oscillator, and the value of its voltage supply. CMOS (complementary metal-oxide semiconductor) is the preferred technology for manufacturing microcontrollers due its low power needs. In static conditions only a very small leakage current flows through the gates. Its power consumption is only significant when switching logic states. Increasing the frequency of the oscillator increases the number of switching actions, and therefore its power consumption also increases. However, it is important to remember that in many applications the microcontroller is just waiting for an external event, such as a key being pressed, or an interrupt, before carrying out a task. Once finished, it returns to the waiting state. To further decrease its power consumption, it is a good idea to paralyze the microcontroller either totally or partially while it is waiting for an external event.

The best method to paralyze the microcontroller is to stop its main oscillator. This will force the main systems to be in a static mode waiting for an external action to start it again. When this happens, the microcontroller is said to be in *idle state, power down,* or *sleep mode*. Different microcontrollers have different methods to enter this low-power state. Some microcontrollers only need to modify a determined bit from a specific register, whereas other microcontrollers have a dedicated instruction for this purpose. The only way to leave this low-power mode is by means of an external interrupt or by a reset.

Example 1.1

8051 microcontrollers have two low-power modes: idle and power down. Any of these two modes can be entered by setting some specific bits of the power control (PCON) registry to 1. In idle mode, the CPU is paralyzed although the main oscillator and the other microcontroller blocks continue working. The microcontroller can leave this mode by means of an external interrupt or a reset. In power-down mode, the oscillator, and therefore the complete microcontroller, become paralyzed. It can only leave the power-down mode by means of a reset.

1.2.4 Protection against Copying

It is important to ensure the safety and protection of the information permanently stored in the microcontroller's memory and to avoid the program to be read or copied from memory once the device has been programmed.

Microcontrollers have resources to protect programs stored in their memory. This protection is normally optional; the programmer has to activate it. Some microcontrollers, like the programmable integrated circuit (PIC) family, can also be configured to prohibit reading of their memory once they have been programmed. Some other microcontrollers have open-memory architecture, that is, they allow the use of memory external to the device. In this case, the protection is done by encrypting the information exchanged by the microcontroller and the external memory. This is typical for the 8051 family of microcontrollers.

Example 1.2

Program protection in 8051 microcontrollers. 8051 microcontrollers have open-memory architectures, allowing the use of external memory. These microcontrollers have two levels of program protection:

Level 1: The stored information is encrypted with an encryption word that can vary between 16 and 64 bits. The encryption is carried out using an XNOR operation between the encryption word and the program in memory. When the CPU reads the content in memory, it carries out another XNOR operation with one of the encryption bits, thus recovering the original bit. This makes it practically impossible to know the real information stored in memory if the encryption word is unknown.

Level 2: A special registry in the microcontroller has security bits that can be programmed to limit total or partial access to the internal program memory.

1.3 Von Neumann and Harvard Architectures

The memory of a microcomputer, microprocessor, or microcontroller stores both data and instructions. Instructions need to move sequentially through the CPU to be decoded and executed. Data can be read from memory by the CPU or written in memory by the CPU. Therefore, the way that memory is organized and the way it communicates with the CPU determines the performance of the device. The two generic hardware models for memory structure are called Von Neumann and Harvard architectures.

Von Neumann architecture was proposed by the mathematician John von Neumann when he designed the *Electronic Numerical Integrator and Calculator* (ENIAC) at the University of Pennsylvania during World War II. He had the seminal idea of developing a stored-program computer. Harvard architecture was proposed by Howard Aiken when he developed the computers known as Mark I, II, III, and IV at Harvard University. These were the first computers to utilize different memories to store data and instructions separately, thus being a much different approach than the stored-program computer.

Figure 1.5 shows these two models. The von Neumann architecture uses a single memory to store instructions and data. This means that one unique address bus can access program instructions and data. Also, a unique data bus can transmit program instructions and data. The CPU

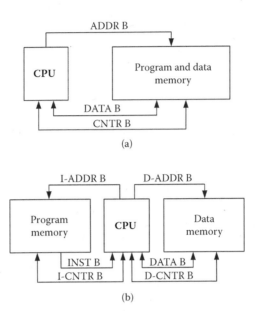

FIGURE 1.5
(a) von Neumann and (b) Harvard architectures. The von Neumann architecture uses a single memory connected to the CPU by using a single address bus (ADDR B), a single data bus (DATA B), and a single control bus (CNTR B). Harvard architecture uses different memories for data and instructions connected to the CPU by an instruction address bus (I-ADDR B), a data address bus (D-ADDR B), an instruction bus (INST B), a data bus (DATA B), an instruction control bus (I-CNTR B), and a data control bus (D-CNTR B).

sends the same control signal to read data or to read an instruction. There are no independent data or instructions control signals. Although ROM is used for instruction storage and RAM is used for data storage, the CPU is not concerned with this distinction and treats them the same way. From the CPU point of view, both ROM and RAM make up a single memory block to which the CPU sends control signals for addresses and data.

Harvard architecture uses different memories to store instructions and data. The program memory has its own address bus (instruction address bus), its own data bus (more properly called an instruction bus), and its own control bus. Data memory has its own address bus, data bus, and control bus independent from the instruction buses. The program memory can only be read when data memory can be read and written.

The von Neumann architecture uses fewer lines than the Harvard architecture, thus making a much simpler connection between CPU and memory. However, this structure does not allow simultaneous handling of data and instructions because there is only one bus. On the other hand, because it has different buses, Harvard architecture allows the handling of data and instructions simultaneously. This gives Harvard architecture an advantage in the speed of execution of programs.

In a microcomputer, the CPU is the microprocessor chip. Because it combines data and program in a single memory, a CPU implemented with the von Neumann architecture will need fewer pins and therefore will reduce the size of the CPU. For this reason, almost all microcomputers using a microprocessor have been developed using the von Neumann architecture. However, the situation is different in a microcontroller. In a microcontroller, the system components are located inside the same integrated chip and therefore there is no need to minimize pins. For this reason, Harvard architecture has been the chosen architecture for most microcontrollers, including the PIC family.

1.4 CISC and RISC Architectures

Complex set instruction computer (CISC) and *reduced instruction set computer* (RISC) are two different computer models classified according to their set of instructions. A CISC has a complex instruction set, whereas a RISC has a reduced instruction set.

When microprocessors and microcontrollers first appeared, the general trend was to give them the most powerful instruction set possible. Therefore, the CISC architecture became the prevalent mode. As time went on, the instructions increased in complexity to the point that the instruction set was a combination of very simple instructions (moving data from memory to the accumulator, for example), and very complex instructions, such as moving a chain of data between memory locations. The length of the instructions was different, the addressing mode became more complex, and in turn all this increased the complexity of the CPU and its size in the chip.

The CPU in RISC architectures has a short set of simple instructions. Each instruction carries out a very simple task (for example, moving data between CPU and memory), but it can be done very fast. Also, all the instructions have the same length. There are few addressing modes and all of them can be applied to any cell. This means that the CPU will be less complex, resulting in it being possible to increase the frequency of the oscillator in order to increase the speed at which operations are executed. Furthermore, as the CPU contains fewer transistors, they are less expensive to design and manufacture. CISC architecture has been the chosen mode for microprocessors and microcontrollers designed since the 1980s. PIC microcontrollers have RISC architecture.

1.5 Manufacturers of Microcontrollers and Microprocessors

Different microcontrollers that have the same core, that is, that share the same CPU and execute the same instruction set, are called a *family of microcontrollers*. Different devices within a family have the same core, but they differ in their I/O capabilities and their memory size. For example, all the microcontrollers in the 8051 family (MCS51) have a similar CPU and execute the same set of instructions. However, different family members have different numbers and types of I/O ports and also different memory types and sizes.

Microprocessors and microcontrollers are manufactured as stand-alone devices—chips that only contain the microprocessor or microcontroller. However, they can also be an embedded-processor core within a large density integration chip that the user will ultimately configure for a particular use. Programmable logic devices (PLD) such as field programmable gate arrays (FPGAs) are an example of such application. PLDs and FPGAs are large integration density circuits in which a user can select their function by choosing the appropriate interconnection elements. One of these elements may be the core of a microprocessor or a microcontroller that the user can connect to part of the memory and the chosen I/O devices. This allows the development of a custom microcontroller for a specific application, while having the advantage that this custom device is compatible with a standard device such as a PIC or 8051 as they both share the same core.

Several industries manufacture microcontrollers and microprocessors in any of the methods discussed earlier. The following is a list of microcontroller and microprocessor manufacturers, as well as of other devices that use a similar common core.

- Actel. FPGA with 8051 and ARM7 cores.
- Advanced Micro Devices (AMD). Microprocessors compatible with xx86.
- Altera. FPGA with Nios II core.
- Analog Devices. Architectures for digital signal processing based on 8052, ARM7, and other processors.
- Applied Micro Circuits Corp. (AMCC). Architectures based on the PowerPC microprocessor.
- ARC International. Architectures based on ARC 600, ARC 700, etc., microprocessors.
- ARM. Architectures based on ARM7, ARM9, ARM10, etc., microprocessor cores.
- Atmel. Architectures based on Marc 4, AVR, 8051, ARM7, ARM9, ARM11, PowerPC, and SPARC.

- Broadcom. Processors for communications and data networks with MIPS architecture.
- Cambridge Consultants. Architectures based on XAP1, XAP2, and XAP3 core processors.
- Cavium Networks. Architectures based on MIPS.
- Cirrus Logic. Architectures based on ARM.
- Cradle Technologies. Digital signal processors: CT3400 and CT3600.
- Cyan Technology. Microcontroller eCOG1k.
- Cybernetic Micro Systems. ASICs with microcontroller P-51.
- Cypress Microsystems. Devices with PSoC (Programmable System-on-Chip) architecture.
- Dallas Semiconductor. 8051-compatible microcontrollers.
- EM Microelectronics. Very low consumption EM6812.
- Freescale Semiconductor (from Motorola). Microcontrollers 68HC05, 68HC08, 68HC11, 68HC12, and 68HC16. DSPs. Processors ColdFire and PowerQuicc with PowerPC core.
- Fujitsu Microelectronics America. Microcontrollers FR80, MB9140x, and F2MC-8FX.
- Goal Semiconductor. Architectures based on 8051.
- Holtek Semiconductor. Microcontroller HT8.
- Hyperstone. Digital Signal Processors E1-32XSR/XSRU, HyNet32S, etc.
- Infineon Technologies (formerly Siemens). Microcontrollers C500, C800, C166, TriCore, etc.
- Infrant Technologies. Microcontrollers for data networks.
- Integrated Device Technology (IDT). Data Communications processors based on MIPS architecture.
- Intel. Microcontrollers from families MCS51, MCS151, MCS251, MCS96, MCS296, etc. Microprocessors xx86, IXP4xx, etc.
- Microchip Technology. Microcontrollers PIC (PICmicro) and digital signal controllers dsPIC.
- MIPS Technologies. Processors MIPS (Microprocessor without Interlocked Pipeline Stages).
- National Semiconductor. Microcontrollers COP8 and CR16, and microprocessors NS32000.
- NEC Electronics America. Microcontrollers 78K0, V850, and others.
- NetSilicon. Processors based on ARM7 and ARM9 cores.
- NXP Semiconductors (formerly Philips Semiconductors). Microcontrollers with 8051, ARM7, and ARM9 cores.

- Oki Semiconductor. Microcontrollers with ARM core.
- PMC-Sierra. MIPS-based processors.
- Rabbit Semiconductor. Processors Rabbit 2000 and 3000.
- Renesas Technology (formerly Hitachi). Microcontrollers R8, H8, and others.
- Sharp Microelectronics. Microcontrollers BlueStreak with ARM7 and ARM9 core.
- Silicon Laboratories. Microcontrollers with 8051 core.
- Silicon Storage Technology. Microcontrollers with 8051 core.
- STMicroelectronics. Microcontrollers with 8051 and ARM7 cores.
- Texas Instruments (TI). Digital signal processors TMS370 and TMS470. Microcontrollers MSP430.
- Toshiba America Electronic Components. Microcontrollers CISC and RISC.
- Ubicom. Microcontrollers SX, IP2000, and IP3000.
- Xemics. Microcontrollers with CoolRISC core.
- Xilinx. FPGA with PowerPC cores.
- ZiLOG. 8-bit microcontrollers with Z8 and Z80 architectures.

2

PIC Microcontrollers

This chapter provides an overview of programmable integrated circuit (PIC) microcontrollers. The chapter starts by describing the general architecture common to the different PIC families, with a special emphasis on several elements based on the working register. It continues with the description of how instructions are executed, the different types of oscillators, the low-power consumption mode, and the watchdog timer. The chapter finishes by discussing the different types of PIC microcontrollers available on the market.

2.1 Main Characteristics of PIC Microcontrollers

All PIC microcontrollers are based on the Harvard architecture as shown in figure 1.5b (chapter 1). This architecture is characterized by having different memories for program and for data. As is common to most microcontrollers, the size of the program memory is larger than the size of data memory. The program memory is organized in words of 12, 14, or 16 bits; the data memory is based on registers of 8 bits. The access to the diverse I/O devices is carried out through some registers in the data memory called *special function registers* (SFRs). Several PIC microcontrollers also have some additional EEPROM to store data in a non-volatile mode.

All PIC microcontrollers are RISC microcontrollers, thus having a relatively reduced number of instructions: between 33 and 77. All the instructions in a PIC family have the same size: 12, 14, or 16 bits. From the programmer's point of view, PIC microcontrollers have a working (W) register and multiple data memory registers. When carrying out arithmetic or logic operations, one of the operands must be in the W register. The resulting value will be placed either in the W register or in any other register in the data memory. Data transfer occurs between the W register and any other register in the data memory, although some high-end PICs allow data transfer directly between two data memory registers. PICs also have instructions to access any bit in any data memory register.

All PIC microcontrollers use pipelining to execute instructions. This pipelining consists of two steps, making up a single instruction cycle.

All instructions, with the exception of control transfer instructions that use two instruction cycles, are executed in a single instruction cycle. An instruction cycle lasts four pulses from the main oscillator.

Another special characteristic of PIC microcontrollers is the implementation of the stack. Here, the stack is not part of the data memory but it has its own independent space, and therefore a finite size. The size of the stack depends on each PIC model. PIC microcontrollers do not have a stack pointer (SP), as is common to most microprocessors and microcontrollers.

PIC microcontrollers have a large variety of I/O devices. They have 8-bit parallel ports, timers, synchronous and asynchronous serial ports, A/D and D/A converters, pulse width modulators, and so forth. The I/O devices generate interrupt requests from the microcontroller. The lower end PICs, however, do not have interrupt resources.

All PIC microcontrollers have a counter that works as a watchdog timer. This timer can be configured with specific bits when the microcontroller is being programmed. Other configuration bits are used to protect the program memory against unauthorized copies.

Many PIC microcontrollers can be programmed in the same circuit for their application with a technique known as *in-circuit serial programming* (ICSP). ICSP uses a small number of lines and therefore it is advantageous.

2.1.1 The Arithmetic and Logic Unit (ALU) and the Working Register in PIC Microcontrollers

The arithmetic and logic unit (ALU) is one of the fundamental components in a microcontroller. The ALU executes the arithmetic and logic operations available in the instruction set. There is one register associated with the ALU that temporarily stores at least one operand involved in the operation, as well as the result of that operation. The ALU also has bits to indicate specific characteristics of the resulting value, such as if the result is zero, the sign of the resulting value, or the existence of carry over. These bits are normally part of the STATUS register.

In most microprocessors and microcontrollers the register associated with the ALU is called the accumulator (ACC). In PIC microcontrollers the register associated with the ALU is called the W register. The W register carries out tasks similar to the ACC, but, as shown in figure 2.1, it is positioned in a different place. Therefore, the ACC and the W register do not operate in the same way.

In traditional architectures, the ACC is placed at the output of the ALU, so it always stores the result of an arithmetic or logic operation. In PIC microcontrollers, however, the result of an operation can either be placed in the W register or in any register in the data memory. This gives PIC microcontrollers an increased amount of computing flexibility and power.

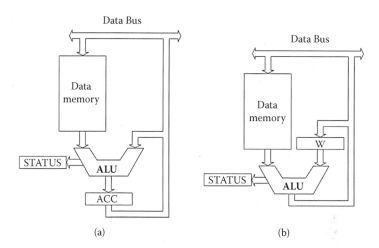

FIGURE 2.1

Relationship between the ALU, working (W) register, and data memory. (a) Configuration used in most microprocessors. (b) Configuration used in PIC microcontrollers. They differ in the location of the W register. This register is called the accumulator (ACC) in microprocessor configurations and the W register in PIC microcontrollers.

2.1.2 Machine Cycles and Execution of Instructions

Like any microcontroller, PIC microcontrollers have a main oscillator to synchronize its internal operations. The pulses from this oscillator (OSC1) are internally divided to generate four signals called Q1, Q2, Q3, and Q4. These signals synchronize all the operations internal to the microcontroller. A machine cycle (MC) is defined as four pulses from the main oscillator (OSC1). Figure 2.2 shows the relationship between OSC1; the Q1, Q2, Q3, and Q4 signals; and the machine cycle.

During the time Q1 is active, in any given machine cycle, the program counter is increased, pointing toward the next instruction to be fetched. This instruction will be fetched during Q4. At the same time, the previous instruction is being executed during the whole machine cycle, that is, from Q1 to Q4.

There are three phases in the execution of a program instruction: fetch, decode, and execution. During the fetching phase, the microcontroller reads the instruction in the program memory and brings it to the CPU. During the decoding phase, the CPU determines the operation to carry out as described by the instruction. Finally, the operation is executed during the execution phase.

Executing an instruction takes two machine cycles. The first cycle fetches the instruction from the program memory. The second cycle decodes and executes the instruction. However, due to the pipelining of operations, the second machine cycle overlaps with the first machine cycle from the next

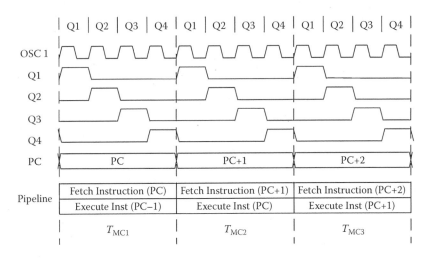

FIGURE 2.2
Clock signals in PIC microcontrollers. OSC1 is the main oscillator from which the internal signals Q1, Q2, Q3, and Q4 are derived. These signals synchronize fetching, decode, and execute of instructions. T_{MC} is the duration of a machine cycle. It uses four OSC1 pulses.

instruction as shown in Figure 2.2. Therefore, from a practical point of view it is possible to say that instructions are executed in one machine cycle.

2.1.3 Pipelining for Instruction Execution

Pipelining is a technique used to overlap two or more instructions as they are being executed. This introduces some parallelism in the execution of instructions, thus reducing the required execution time. The programmer does not need to worry about pipelining, as it is incorporated into the design of the microcontroller.

Pipelining is similar to a production line in a factory. There, the product is moved between stations, each one of them doing a specific task. In a production line with n steps, there are always n products in the process of being manufactured. Let's assume T_s is the time that the product spends in each station. The total production time will then be $n \times T_s$. But as there is a product coming out from the production line at any T_s, time units, the average time for product manufacturing is then T_s. In an n-stage pipeline, each instruction spends a time equal to T_{MC} for any stage, with T_{MC} being the time length of a machine cycle. Therefore, the time needed to move through all the stages is $n \times T_{MC}$. However, because instructions exit the pipeline every T_{MC} seconds, it is possible to assume that the average time to execute any instruction is T_{MC}. Because some instructions, such as control transfer instructions, require additional machine cycles, the average instruction execution time for these instructions is slightly longer than T_{MC}.

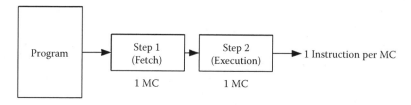

FIGURE 2.3

Two-step pipelining. The first step fetches the instruction that will be executed in the second step. Each step lasts one machine cycle (MC). The pipeline has two different instructions in the two steps. Each instruction stays in the pipeline for 2 MC, therefore there is one instruction leaving the pipeline each MC.

Figure 2.3 shows how the PIC microcontroller executes instructions in a two-stage pipeline. Each instruction requires two stages. The first stage is the fetching stage that requires a machine cycle. The second stage, in which the instruction is decoded and executed, requires another machine cycle. Therefore, during each machine cycle, the microcontroller fetches one instruction and executes the previous instruction. Every machine cycle period results in an instruction being executed, with the previously mentioned exception of control transfer instructions as described in Example 2.1:

Example 2.1

Figure 2.4 shows the operation of a two-stage pipeline when executing a program segment that includes a control transfer instruction. During the first machine cycle (MC1), the microcontroller fetches instruction I1 while executing I0 (not shown in the figure). During MC2, it fetches I2 while executing I1. During MC3, the microcontroller fetches I3, a control transfer instruction, while it is executing I2. During MC4, it fetches I4 while executing I3. I3 is a subroutine call that starts at instruction I10. It puts the current program counter in the stack and points the program counter toward instruction I10 that will be executed at the next MC. At MC5, the microprocessor fetches I10, but I4 that was already in the pipeline must not be executed. I4 is taken away from the pipeline and replaced by a no-operation (nop) instruction. At MC6 the next instruction (I11, not shown in the figure) is fetched while I10 is executed. It is then possible to see how the control transfer instruction needs two machine cycles.

2.1.4 Oscillators

The main oscillator in PIC microcontrollers can be a crystal oscillator, an *RC* oscillator, or an external clock. Some devices also have an internal *RC* oscillator at 4 MHz. Increasing the oscillator frequency shortens the length of the machine cycles and therefore the time needed for executing instructions, but also increases power consumption. The type of oscillator can be

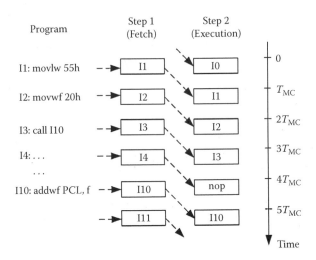

FIGURE 2.4

Example of instruction flow using two-step pipeline. Instruction I3, being a control transfer instruction needs two machine cycles (MCs) to be executed. The rest of the instructions are executed in a single MC. T_{MC} is the length of an MC.

selected by the configuration bits. These also select specific modes of operation for crystal or ceramic oscillators: LP, XT, and HS. The LP mode selects oscillator frequencies between 32 kHz and 200 kHz, and is used in very low power-consumption applications. The XT mode selects oscillator frequencies between 100 kHz and 4 MHz. The HS mode selects oscillator frequencies between 8 MHz and 20 MHz. Figure 2.5 shows a general configuration for a crystal oscillator.

The *RC* oscillator is the least expensive option for the main oscillator in the microcontroller. This can only be used when frequency accuracy and stability are not critical. Figure 2.6 shows how this oscillator is implemented in a medium-end PIC. Although manufacturers do not have an equation to determine the values of C_{EXT} or R_{EXT}, they provide graphs that relate the value of the oscillation frequency with V_{DD}, C_{EXT}, and R_{EXT} at 25°C.

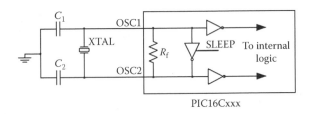

FIGURE 2.5

Crystal oscillator. C_1, C_2 = 15 pF to 68 pF for a 4 MHz crystal (XTAL).

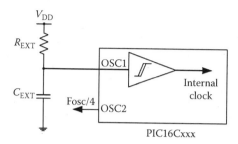

FIGURE 2.6
RC oscillator. $R_{EXT} = 3$ kΩ to 100 kΩ; $C_{EXT} > 20$ pF.

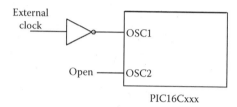

FIGURE 2.7
Using an external clock as an oscillator in a PIC.

The last option is to use an external clock as the main oscillator for the microcontroller. Figure 2.7 shows this option for a medium-end PIC microcontroller.

2.1.5 Configuration Bits

All PIC microcontrollers have specific bits for their configuration. These configuration bits are stored in non-volatile memory (EEPROM) when the device is being programmed. However, they are not accessible once the PIC is executing the program. More specifically, the program cannot modify these bits. The configuration bits allow the programmer to specify certain aspects of the microcontroller to better fit the intended application. Configuration bits can modify the following parameters:

- Type of oscillator
- Watchdog timer on or off
- Program memory protection
- Protection of EEPROM data if available
- Specifications for reset and power supply

Not all these parameters can be programmed in all devices. The following example describes how these bits are programmed.

Example 2.2

Configuration bits in a medium-end PIC make up a word stored in address 2007h in program memory. This address is only accessible during the programming of the microcontroller. Once the program is being executed, it cannot access this address. Figure 2.8 shows the configuration bits for the PIC16F873.

2.1.6 Reset Options

Reset sets the microcontroller in a known, predetermined state. During the transient time of reset, the microcontroller is temporarily paralyzed and not executing any program instructions. Following this transient state, the device moves to the known, predetermined state. In PIC microcontrollers, reset sets the program counter to zero. Therefore, after the

13	12	11	10	9	8	7	6	5	4	3	2	1	0
CP1	CP0	DEBUG	–	WRT	CPD	LVP	BODEN	CP1	CP0	PWRTE#	WDTE	FOSC1	FOSC0

CP1, CP0: FLASH program memory code protection:
      11 - Code protection off
      10 - Only last 256 cells (F00h to FFFh) protected
      01 - Only page 1 (800h to FFFh) protected
      00 - All memory protected (000h to FFFh)
DEBUG: In-circuit Debugger mode
      1 - disabled (RB6 and RB7 are general purpose I/O pins)
      0 - enabled (RB6 and RB7 are dedicated to the debugger)
WRT: FLASH program memory write enable:
      1 - enabled
      0 - disabled
CPD: Data EEPROM memory code protection:
      1 - Code protection off
      0 - Code protection on
LVP: Low voltage In-circuit serial programming:
      1 - enabled
      0 - disabled
BODEN: Brown-out reset:
      1 - enabled
      0 - disabled
PWRTE#: Power-up timer:
      1 - disabled
      0 - enabled
WDTE: Watchdog timer:
      1 - enabled
      0 - disabled
FOSC1, FOS0: Oscillator selection bits:
      11 - RC
      10 - HS
      01 - XT
      00 - LP

FIGURE 2.8
Configuration bits for PIC16F873.

reset has finished, the first instruction executed is the one located at this memory address, independently of what happened before the reset.

There are several reasons that can originate a reset; these are known as *reset sources*. Reset sources can be different for different microcontrollers. The following are common to most PIC microcontrollers:

- External reset
- Power-on reset
- Watchdog reset
- Brown-out reset

Figure 2.9 shows the logic circuit used to produce the reset signal in PIC microcontrollers. The circuit needs to ensure that once the reset transient is finished, the microcontroller is in a stable state. More specifically, this circuit needs to guarantee that the microcontroller will only leave the reset state if the voltage has reached a stable and high enough value. This is the task of the blocks associated with the V_{DD} signal in Figure 2.9. These blocks also function during a brown-out reset.

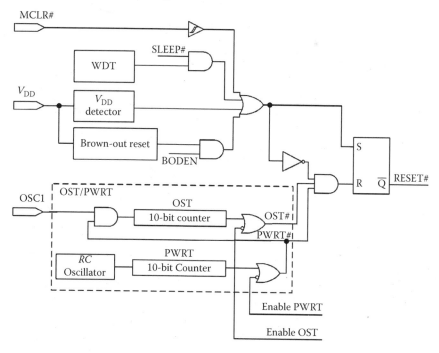

FIGURE 2.9

Simplified block diagram for reset in a PIC microcontroller. The operation of some of these blocks can be programmed through configuration bits (BODEN, PWRTE, etc.).

Furthermore, the reset circuit has to guarantee that the microcontroller will only leave the reset state if the main oscillator is working and is stable. This is the task of the block labeled OST/PWRT in Figure 2.9. It takes a certain amount of time for the main oscillator to reach stable values for its frequency and amplitude after it has been turned on. The microcontroller should not leave the reset state if frequency and amplitude are not yet stable. The main oscillator is turned on when the microcontroller is first powered, when it leaves its low-power consumption mode, or in the case of a brown-out. These cases correspond to the power-on reset, reset due to low-power consumption mode, and brown-out reset.

The block labeled OST/PWRT in Figure 2.9 has two timers: oscillator start-up timer (OST) and power-up timer (PWRT). An internal RC oscillator, independent of the main oscillator, introduces a 72 ms delay to the signal PWRT after the oscillator has been powered. The OST introduces an additional delay of 1024 pulses. This delay is long enough for the oscillator to reach stable amplitude and frequency values. The OST starts its operation only after the PWRT has reached its maximum value.

Figure 2.10 shows the time sequence for the signals associated to the OST/PWRT block in two situations: during a power-on reset and during a manual reset.

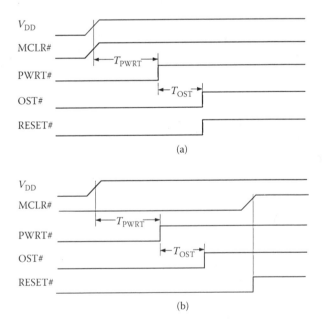

FIGURE 2.10
Time diagrams showing the sequence of signals associated with the OST/PWRT block. (a) Power-on reset—MCLR# pin connected to V_{DD}. (b) Manual reset.

An external reset occurs when the pin MCLR# is set to 0. MCLR# must be at logic value 1 during the normal operation of the microcontroller. An external reset can occur during the regular operation of the microcontroller or when the microcontroller is in a low-power mode (SLEEP). It is possible to connect an external switch to pin MCLR# to create a manual reset as shown in Figure 1.4a.

Power-on reset occurs when MCLR# is connected to the microcontroller's power supply as shown in Figure 2.11. Once the microcontroller detects the triggering of V_{DD} it creates a reset signal to guarantee that the microcontroller will start operating correctly. This configuration, connecting MCLR# to V_{DD} directly or using a resistor, as shown in Figure 2.11a, avoids the need to introduce external circuits. If the power supply has a long settling time, it is necessary to guarantee that the voltage at MCLR# will be below the threshold until V_{DD} reaches its appropriate value. This is shown in Figure 2.11b.

A watchdog reset occurs when the watchdog timer reaches its maximum value. This will happen when the program running in the microcontroller cannot clear the internal counter of the watchdog time before reaching its maximum value. A watchdog reset may happen when the microcontroller is in its regular working mode or when it is in the low-power consumption mode. In this last case, the microcontroller will leave the low-power mode without producing a reset. Section 2.1.8 describes in further detail how the watchdog timer operates.

Brown-out reset occurs due to a transient or glitch in the voltage (V_{DD}) supplied to the microcontroller. The microcontroller has a circuit that will produce a reset signal in this situation and will keep the microcontroller in this state while the voltage V_{DD} is below a predetermined threshold as was shown in Figure 2.9. Once V_{DD} recovers above the threshold, the PWRT keeps

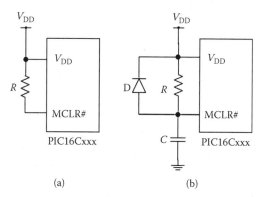

(a)   (b)

FIGURE 2.11
(a) Circuit to guarantee power-on reset: MCLR# and V_{DD} pins can be connected directly or through a resistor. (b) Circuit for power-on when the voltage supply has a long settling time.

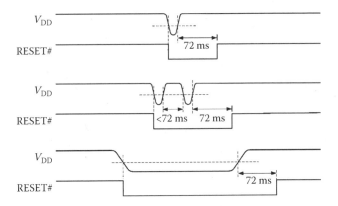

FIGURE 2.12
Reset signal due to brown-out in several situations. The timer PWRT guarantees that V_{DD} will be at its nominal value once it leaves the reset state.

the reset signal for an additional 72 ms. This guarantees that the main oscillator and V_{DD} are within their nominal parameters after leaving the reset state. Figure 2.12 shows different situations related to brown-out reset.

When programming the microcontroller it is also possible to configure some of its reset sources. More important, however, is that after a reset has occurred it is possible to find out the origin of the reset signal. This can be done by reading some special function registers such as STATUS and PCON.

Example 2.3

Figure 2.13 shows the PCON register for the PIC16F873 that is a medium-end microcontroller. Two bits in this register are used to determine if the reset signal was a power-on reset (POR) or a brown-out reset (BOR) by setting the appropriate bits to 0. The program must set both bits (POR# and BOR#) to 1 after the reset. Also, Figure 3.14 (Chapter 3) shows the register STATUS in which the bit TO# is set to 0 when the watchdog timer produces a reset.

PCON

POR#: Power-on reset indicator
      1 – Power-on reset did not occur
      0 – Power-on reset occurred
BOR#: Brown-out reset indicator
      1 – Brown-out reset did not occur
      0 – Brown-out reset occurred

FIGURE 2.13
Two bits in the special function registry PCON in PIC16F873 are used to determine the origin of a reset signal.

2.1.7 Low-Power Consumption Mode

When the microcontroller is in low-power consumption mode (*sleep mode*), most of its functions, including the main oscillator, are stopped. The power consumed by the microcontroller in these conditions is extremely low, less than 1 µA for some models.

The instruction sleep places the microcontroller in the low-power consumption. While in this state, the values stored in the data memory registers are not changed. Because of the pipeline process, in which the instruction after sleep has already entered the CPU, this instruction will be executed once the microcontroller wakes up. For this reason it is recommended that the instruction after sleep is an nop instruction. Sleep also sets the watchdog timer counter to zero.

The microcontroller will wake up leaving the low-consumption mode in any of these three events:

- Reset
- Overflow of watchdog timer counter (if not disabled)
- Interrupt, either external or from its internal peripherals

If the microcontroller wakes up due to a reset, it will execute the instruction stored at address 0 in the program memory. If it wakes up due to the watchdog timer, it will continue running the program, executing the instruction after the sleep instruction. If the microcontroller wakes up due to an external interrupt with the interrupt system enabled, it will execute the instruction immediately after the sleep instruction, and then the program counter will jump to address 4 in the program memory searching for the interrupt routine. However, if the interrupt system is not enabled, the microcontroller wakes up, executes the instruction after the sleep instruction and continues with the program sequence without jumping to address 4. The interrupt system can be enabled or disabled by modifying the global interrupt enable (GIE) bit. When GIE equals 1 the interrupt system is enabled; when GIE equals 0 it is disabled. GIE is bit number 7 in the special function register INTCON.

2.1.8 Watchdog Timer

The watchdog timer (WDT) consists of an oscillator and a pulse counter. The WDT (watchdog timer enable) oscillator is independent from the main oscillator; it continues working when the microcontroller is in low-power mode. If the pulse counter reaches its maximum value (overflows) during the normal operation of the microcontroller, the WDT times out and generates a reset signal for the microcontroller. If the pulse counter reaches its maximum value when the microcontroller is in a low-power

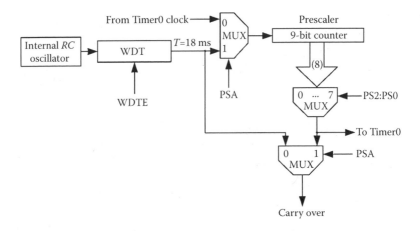

FIGURE 2.14
Block diagram of the circuits associated with a watchdog timer in medium-end PICs. PSA and PS2:PS0 are bits from the special function registry OPTION. WDTE is the configuration bit that enables the watchdog timer.

mode, the microcontroller wakes up executing the instruction right after the sleep instruction.

Figure 2.14 shows the block diagram for the WDT in a medium-end PIC. Bit WDTE (watchdog timer enable) enables the watchdog timer. Once the program is running normally, the WDT cannot be disabled.

The WDT times out every 18 ms. Therefore, to avoid the WDT timing out and generating the reset signal for the microcontroller, the internal pulse counter needs to be set to 0 before the 18 ms have elapsed since the last time it was set to 0. The instruction clrwdt is used for this purpose. It is possible to extend the 18 ms timeout up to 2.3 s by assigning an additional counter to the WDT. This counter is called prescaler. Table 2.1 shows the available time out times for the four configuration bits (PSA, PS0, PS1, and PS2) of the prescaler. These bits belong to the register OPTION. PSA must be at value 1 to assign this prescaler counter to the WDT.

2.2 PIC Microcontroller Families

PIC microcontrollers can be classified into three main types according to the length of their instructions:

- Low-end microcontrollers: 12-bit instructions
- Medium-end microcontrollers: 14-bit instructions
- High-end microcontrollers: 16-bit instructions

TABLE 2.1

Division Factors for a Prescaler and Their Effect on Watchdog Timer
Overflow

PS2:PS0	Prescaler Division Factor	Approximate Watchdog Timer Overflow (ms)
000	1:1	18
001	1:2	36
010	1:4	72
011	1:8	144
100	1:16	288
101	1:32	576
110	1:64	1152
111	1:128	2304

Note: PS2, PS1, and PS0 are bits from the OPTION special function register.

An alternative method to classify PIC microcontrollers is by their number of pins. This way, PIC microcontrollers can be classified as PIC10, PIC12, PIC16, PIC17, and PIC18. Some of these families, such as the PIC16, have large subfamilies. Furthermore, the devices in each one of these families may have instructions with different lengths. This is the case of the PIC12 and PIC16 families that have microcontrollers in the low- and medium-end ranges. Table 2.2 shows these two classification modes.

2.2.1 Low-End Microcontrollers

Low-end PIC microcontrollers have an instruction set of 33 instructions, each one being 12 bits long. Program memory can be up to 2048 words, with each word also having 12 bits. This memory is organized in pages of

TABLE 2.2

Summary of PIC Microcontroller Families

Family	Low-end	Medium-end	High-end	Main Characteristic
PIC10	X			6 pins
PIC12X5	X			8 pins
PIC12 (except PIC12X5)		X		8 pins
PIC16X5	X			—
PIC16 (except PIC16X5)		X		—
PIC17			X	—
PIC18			X	Improved high-end

512 words. Data memory consists of 8 bit registers, organized in banks of up to 32 registers.

Low-end PICs have a two-level stack to store program memory addresses. These PICs do not allow interrupts. Their I/O resources have a low number of devices, up to three input/output ports each one of 8 bits, a timer, and a comparator.

Low-end PICs can be further classified as:

- PIC16X5xx family
- PIC12X5xx family
- PIC10Fxxx family

PIC16X5xx can be considered as the main family of low-end PIC microcontrollers. Their program memory can be EPROM, OTP, or flash depending on the model. Their power consumption is less than 2 mA at 5 V in normal operating mode, and less than 3 µA at 3 V for low-consumption mode. They are available with 10, 20, or 28 pins. Figure 2.15 shows the internal architecture of the PIC16X5xx family.

The PIC12X5xx family is characterized by using an eight-pin package. Given the low number of pins, their I/O resources are limited to a 6-bit parallel port, a timer, and an A/D converter depending on the specific model. The program memory can be OTP or flash. Some models have an EEPROM for data memory. Power consumption is less than 2 mA at 5 V in regular operating mode, and less than 2 µA at 3 V in low-power mode. Figure 2.16 shows their internal architecture.

The PIC10Fxxx microcontrollers are quite small and are available in packages of six or eight pins. Their program memory is flash, although they do not have additional EEPROM for permanent data storage. Their I/O resources are limited to a parallel port of 4 bits, a timer, and a comparator. Power consumption is less than 350 µA at 2 V in regular operating mode, and less than 100 nA at 3 V in low-power mode. Figure 2.17 shows their internal architecture.

2.2.2 Medium-End Microcontrollers

Figure 2.18 shows the generic architecture of medium-range PICs. These have an instruction set of 35 instructions, each one 14 bits long. Their program memory can be up to 8192 words, each one also of 14 bits. This program memory is organized in pages of 2048 words. Their data memory is made up of 8-bit registers, organized in banks of 120 registers. Medium-end PICs can have up to four banks. In general, medium-range microcontrollers have some EEPROM data memory and an eight-level stack to store program memory addresses.

These PICs have a system of fixed interrupts for internal interrupts (from their internal systems) and one external interrupt. Each I/O block can

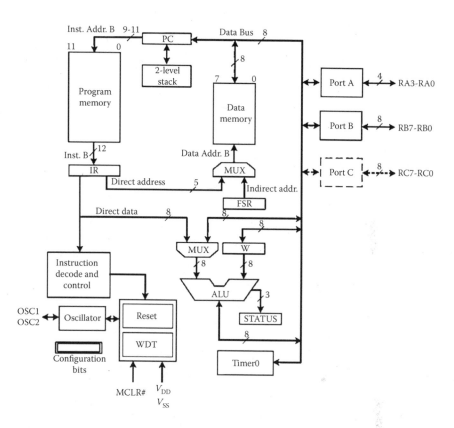

FIGURE 2.15
Internal architecture of PIC16X5xx family.

generate an interrupt request to the CPU. All medium-range PICs have a terminal to receive interrupt requests from external devices. Their I/O devices are several parallel ports (ports A, B, C, etc.); up to three timers; two modules for capture, comparison, and pulse width modulation; several serial ports for synchronous and asynchronous communication; a 10-bit A/D converter associated to an analog multiplexer, and so forth. Figure 2.19 shows the internal architecture for the PIC16F873, used in several examples in this book.

Medium-range PICs can be further classified as:

- PIC16 (except the PIC16X5xx family that are low-end PICs)
- PIC12X6xx; these have an eight-pin package

Medium-end PICs with an eight-pin package can operate at a low voltage (2 V). Their power consumption is 100 μA for their normal operation, and 1 nA in low-power mode. Figure 2.20 shows the internal architecture for a medium-range PIC with an eight-pin package.

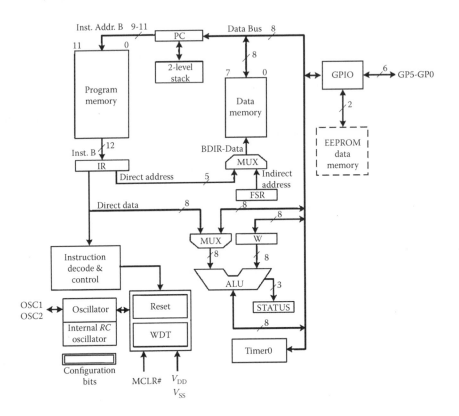

FIGURE 2.16
Internal architecture of PIC12X5xx microcontroller family.

2.2.3 High-End Microcontrollers

High-end microcontrollers are characterized by using 16-bit instructions, a deeper stack, and an interrupt system that can handle internal interrupts as well as several inputs for external interrupts. Some of these PICs have an open architecture. This allows the increase of both program memory and data memory. The number of available devices for high-end microcontrollers is also larger.

High-end microcontrollers can be further classified as:

- PIC17
- PIC18

PIC17 microcontrollers have an instruction set of 58 instructions, each one of 16 bits. Their program memory size can be up to 65,536 words of 16 bits each. Their data memory can be up to 1024 registers of 8 bits. Program memory can be EPROM, ROM, or OTP. The stack has 16 levels. Their interrupt system can handle different priorities.

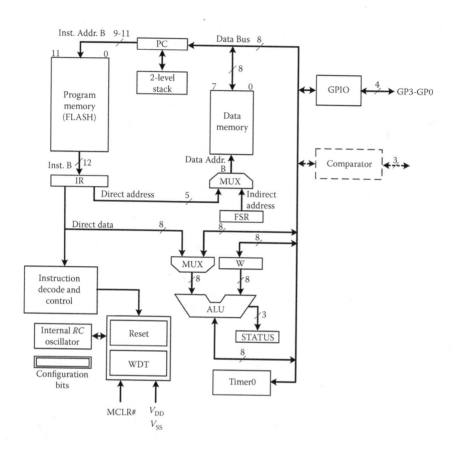

FIGURE 2.17

Internal architecture of the PIC10Fxxx. These are low-end microcontrollers with flash memory and six-pin packaging.

PIC17 microcontrollers have an open architecture. The devices in the PIC17 family can work in four configurations: microcontroller, protected microcontroller, extended microcontroller, or microprocessor. When operating in the microcontroller or protected microcontroller modes, the PIC17 can only access its internal memory. When working as an extended microcontroller or as a microprocessor, it is possible to use external memory that the microcontroller can access.

These PICs have several I/O devices, including parallel ports, serial ports, timers, and A/D converters. Figure 2.21 shows the internal architecture of the PIC17 family.

Microcontrollers that belong to the PIC18 family have flash memory and an instruction set of 77 instructions of 16 bits each. Their program memory can have a size of up to 2 MB, and their data memory can reach 4096 registers of 8 bits in each one. Some members of the PIC18 family allow external memory to store program memory. Their stack is 31 levels deep;

they can handle internal interrupts as well as three external interrupts. Some devices of the PIC18 family have been designed to work with low voltages (from 2.0 V to 3.6 V) using less than 2 mA. Figure 2.22 shows the architecture of the PIC18 family.

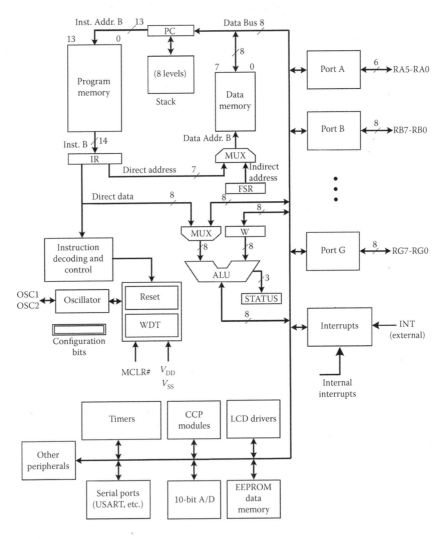

FIGURE 2.18
General architecture of medium-end PIC microcontrollers.

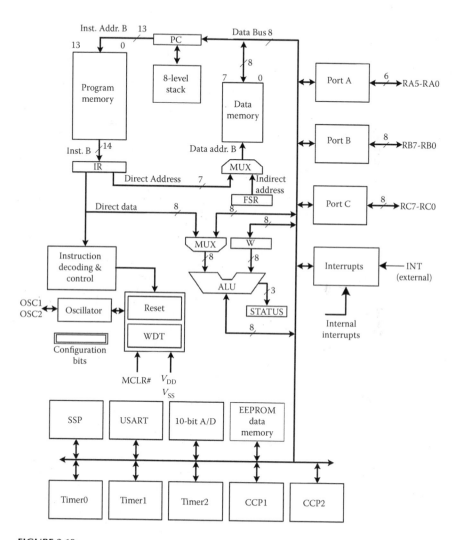

FIGURE 2.19

Internal architecture of PIC16F873, shown as an example of a medium-end PIC microcontroller.

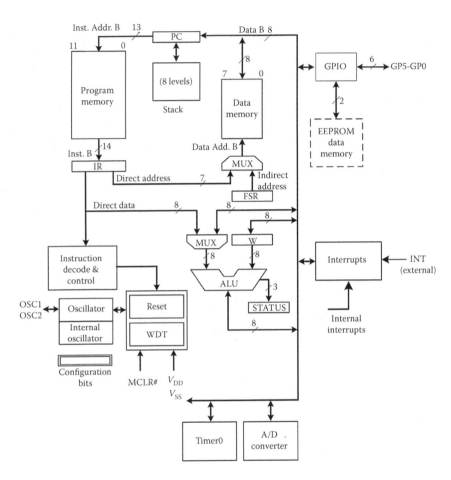

FIGURE 2.20
Internal architecture of PIC12CE67X family. These are medium-end microcontrollers with an eight-pin package.

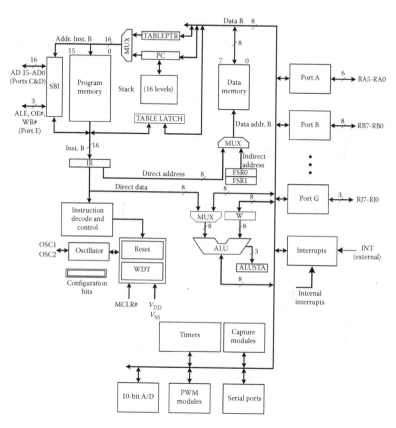

FIGURE 2.21

Internal architecture of PIC17. Their program memory can be externally expanded.

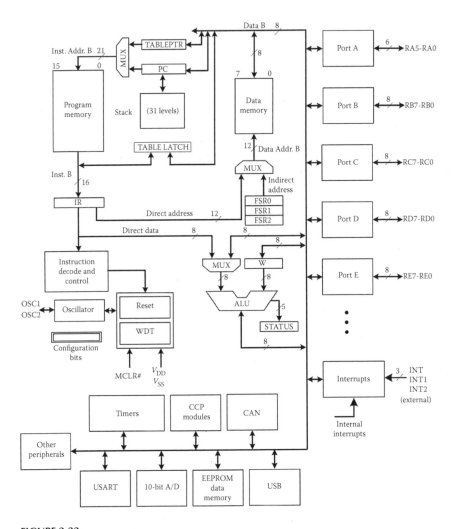

FIGURE 2.22
Internal architecture of PIC18. These microcontrollers have superior performance to medium-end PICs.

3

Memory in Microcontrollers

This chapter describes the general structure and organization of memory in microcontrollers, in particular medium-end PICs. The chapter begins by defining basic memory concepts, including word, address, and memory size. This is followed by the description of the two methods to organize memory in microcontrollers: linear memory and paged memory. The chapter then describes several technologies used for memory fabrication and finishes by explaining how memory is organized in medium-end PIC microcontrollers.

3.1 Basic Concepts

Memory in a microcontroller is the place that stores the program being executed and the data or variables used by that program. Memory can be considered as a set of cells or locations, identified by their address. Each cell stores a word. A *word* is the logic unit of information stored in a cell. Words can have 1, 8, 12, 14, or 16 bits. An 8-bit word is called a *byte* (B; Figure 3.1)

The number of memory cells determines its *size*. Size is measured in words; according to their bits or bytes; or words having 12, 14, or more bits. While the International System of Units established prefixes that are factors of 10 (kilo, mega, giga, tera, etc.), it is customary in computer science to utilize the same prefixes to indicate multiplication factors that are powers of 2. To avoid confusion, in 1998 the International Electrotechnical Commission (IEC) introduced new prefixes to be used for multiplication factors that are powers of 2. Table 3.1 shows these prefixes. However, as these new prefixes are not commonly used in practice, this book will continue using the notation of powers of 10 prefixes. For example, when referring to a memory size of 1024 bytes, we will write 1 kB instead of using 1 KiB, which would the correct way to write it according to the IEC.

The *address* identifies a specific cell in the whole memory. The simplest way of identifying cells is to assign to each one of them an integer number that is continuously increasing. This makes the address equal to the binary number that identifies a specific cell (Figure 3.2). With D being the address of a generic cell, the possible address values in a memory of N cell are

TABLE 3.1

Examples of Prefixes, Symbols, and Multiplication Factors

International System of Units			International Electrotechnical Commission		
Prefix	Symbol	Factor	Prefix	Symbol	Factor
kilo	k	10^3	kibi	Ki	2^{10}
mega	M	10^6	mebi	Mi	2^{20}
giga	G	10^9	gibi	Gi	2^{30}
tera	T	10^{12}	tebi	Ti	2^{40}

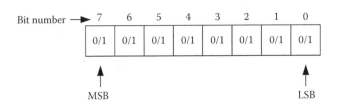

FIGURE 3.1

An 8-bit word (byte), indicating the positions of the most significant bit (MSB) and least significant bit (LSB) within this word.

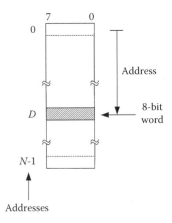

FIGURE 3.2

Addresses in a memory of N cells, with each cell being 8 bits long.

$$D = 0, 1, 2, ..., (N - 1). \tag{3.1}$$

The number of bits (n) needed to identify the address of a cell depends on the memory size (N):

$$N = 2^n. \tag{3.2}$$

For example, a 1 kB memory needs $n = 10$ bits in order to specify the address of any cell as $2^{10} = 1024$, which is the number of different cells in the memory. The allowed memory addresses are $D = 0, 1, 2, ..., 1023$. Using hexadecimal notation, as is customary in computer science, the allowed memory addresses can be written as $D = 0, 1, 2, ..., 3FFh$.

3.1.1 Logic Organization of Memory

There are two main methods of organizing memory in microcontrollers: as a single block (linear organization) or by sets of blocks called *pages*. In linear organization, the cell addresses are consecutive binary numbers. Each cell is identified by its linear address (D), made up by a unique binary number. Linear addresses conform to Equation 3.1.

A memory page is a fixed-size portion of memory. Pages are consecutive and do not overlap. Each page can be identified with a consecutive number called a page number. Inside a page, cells are identified by their position—called *displacement*—relative to the beginning of the page. Within a page-organized memory such as shown in Figure 3.3, the address of a specific cell is a combination of two elements: page number (*pgnum*) and its displacement (*disp*). These two elements make up the logic address (*Logaddr*) for the cell:

$$Logaddr = pgnum : disp. \tag{3.3}$$

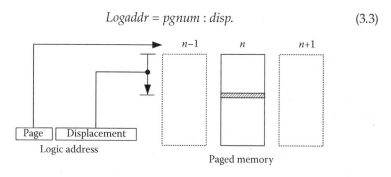

FIGURE 3.3
Memory organized in pages. The logic address for a single cell is made of the page number and the displacement within the page.

Any linear address can be obtained from the logic address by multiplying the page number by the size of the page (*pgsz*) and adding the displacement of the cell within the page:

$$D = pgnum \times pgsz + disp. \tag{3.4}$$

For example, in the memory shown in Figure 3.4 organized in pages of 256 B (100h), a cell located in page 2 with a displacement of A5h has a linear address equal to $D = 2 \times 100h + A5h = 2A5h$.

It is important to note that if the size of the page is a power of 2 (that is, $pgsz = 2^k$), the number needed to indicate the displacement within a page is a binary number of *k* bits. Working in binary, the product *pgnum* × *pgsz* adds k zeros to the right of *pgnum*. Adding *disp* as indicated in Equation 3.4, the value *disp* takes the place of these zeros. Therefore, in the resulting linear address, the most significant bits are *pgnum* and the least significant bits are *disp* as shown in Figure 3.5.

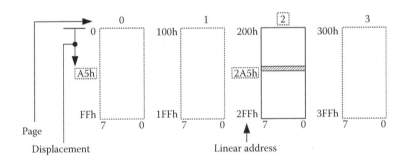

FIGURE 3.4
1 kB memory paged in 256 B pages showing the relationship between the linear address and the logical address. The linear address of the cell located in page 2 with a displacement A5h is 2A5h.

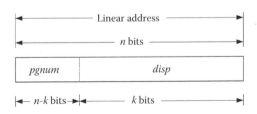

FIGURE 3.5
Linear address based on the components of logic address in a paged system when page size is a power of 2. The displacement (*disp*) takes the *k* least significant bits, and the page number (*pgnum*) takes the *n* − *k* most significant bits in the *n* bit linear address.

3.1.2 Types of Memory

Program memory in a microcontroller is non-volatile, read-only memory. Several technologies can be used for manufacturing program memory: read-only memory (ROM), erasable programmable ROM (EPROM), one-time programmable (OTP), and flash. On the other hand, data memory needs to be read and written, but the information stored does not need to be preserved once the microcontroller has been unpowered, so volatile memory can be used. Data memory normally uses static random-access memory (RAM). Some microcontrollers also use external, non-volatile memory for data memory. This is done by using electrically erasable programmable read-only memory (EEPROM) to store values of data that are fixed or do not change often.

RAM: RAM memory can be read and written. There are two types of RAM: static RAM and dynamic RAM. Information stored in static RAM is kept indefinitely as long as the memory is powered. Dynamic RAM must be refreshed periodically to keep the information stored. Dynamic RAM is widely used in personal computers but not in microcontrollers.

ROM: For those microcontrollers that use ROM memory, the information is written during the manufacturing process and cannot be changed later. For this reason, it is necessary to ensure that the program code and fixed data values are correct before the microcontroller is created. Using ROM is the most economic way of producing large quantities of microcontrollers for a given application. In PIC microcontrollers, the label "CR" indicates that the program memory uses ROM memory, for example, PIC16CR65 and PIC16CR72.

EPROM and OTP: EPROM and OTP memory are very similar; they only differ in their packaging. The packaging for EPROM devices has a glass window that allows ultraviolet light to go through to erase the device. OTP devices use the same internal memory without the glass window. Therefore, once the user has programmed them, they cannot be altered or erased. In PIC microcontrollers, the label "C" indicates that they use EPROM or OTP memory. For example, PIC16C74B/JW is an EPROM device, as the suffix JW indicates the existence of the glass window to erase its contents. The PIC16C72A/P and PIC16C74B/SO microcontrollers use OTP memory. Their packaging is plastic DIP (dual in-line package) for the first one and plastic SOIC (small-outline integrated circuit) for the second one.

EEPROM: EEPROM is non-volatile memory that can be read and written electrically. Cells do not need to be previously erased in order to store new information in them. The number of times that the EEPROM memory can be written is finite, although it is a large value, on the order of millions.

Flash: Information in flash memory can be read and written to individual cells. However, existing data need to be erased before new data can be written in a cell. Erasing occurs in blocks of cells instead of erasing individual cells; this is the main difference compared to EEPROM memory. Erasing data means to set to 0 all the bits in the block of cells. Writing data means to set specific cells to 1. It is possible to set to 1 a bit that was at 0, but in order to set to 0 a bit that was previously at 1, it is first necessary to erase its block of cells. Therefore, writing information in a cell in flash memory becomes an operation involving reading, erasing, and writing the block of cells in which we want to store that information. All these operations (erasing, writing, reading) are carried out using the nominal supply voltage, without it being necessary to use higher voltages. Flash memory can be written or erased a finite number of times, in the range of hundred thousands of times. The label "F" in PIC devices indicates that they use flash memory, for example, PIC16F873.

3.2 Memory in Medium-End PIC Microcontrollers

Memory in PIC microcontrollers is organized according to the Harvard architecture. Therefore, there are two independent memory spaces: one for program and one for data. Program memory is a read-only memory built on ROM, OTP, EPROM, or flash technologies. It stores the program instructions for the microcontroller to execute. Some PIC microcontrollers allow the program to read its own memory, thus making it possible to store data that does not change in the program memory. In some other PIC microcontrollers with flash memory, it is possible to write data in the program memory.

Data memory is built on static RAM. Therefore, it is a volatile, read and write memory. Some PIC models also have an additional EEPROM memory to store data that does not change very often. Table 3.2 shows the size of memory available in some medium-end PIC microcontrollers.

3.2.1 Program Memory

Program memory is organized in pages. Medium-end PICs can have up to four pages of 2k words each, making a total memory size up to 8k words. The length of a word in medium-end PIC microcontrollers is 14

TABLE 3.2

Size of Program and Data Memories for Several Medium-End PIC
Microcontrollers

PIC	Program Memory (in 14-Bit Words) Flash	Data Memory (in Bytes)	
		RAM	EEPROM
PIC18F83	512	36	64
PIC16F84/84A	1024	68	64
PIC16F873/874	4096	192	128
PIC16F876/877	8192	368	256

bits. Figure 3.6 shows the organization of memory. Address 0h in page 0
is reserved for reset, while address 4h in the same page 0 is reserved for
the interrupt vector. These addresses hold the instructions to go to the
appropriate programs.

3.2.1.1 Addressing Program Memory

The program counter (PC) is a register in the PIC microcontroller that
addresses the program memory. In particular, the PC stores the address
of the next instruction to be executed. That is, the PC points toward the
instruction that will be executed after the instruction currently being executed. The PC in medium-end PIC microcontrollers is 13 bits long, thus
allowing for the addressing of 8k words in the program memory. Due to
the paging system, bits 12 and 11 in the PIC indicate page number, and bits
10 to 0 indicate the address within that page.

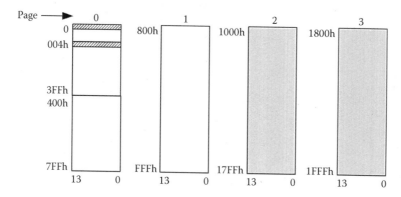

FIGURE 3.6

Program memory pages in medium-end PIC microcontrollers. They can have up to four
pages of 2k words, each one with a total of up to 8k words. Each word is 14 bits long. The
shadowed part is not part of the PIC16F873 because this microcontroller has 4k cells distributed between pages 0 and 1. The PIC16F84 only has the first 1024 memory cells that are
located from address 0 to address 3FFh in page 0.

The PC is related to two special function registers in the data memory: PC latch high (PCLATH) and PC low (PCL). The 8 least significant bits in the PC are the PCL register. This register can be read and written by the programmer. The 5 most significant bits in the PC (PCH) cannot be read, although they can be modified through the PCLATH register. As shown in Figure 3.7, the page number is loaded in bits 4 and 3 in PCLATH (PCLATH <4:3>). Therefore, it is possible to know at any time the content of the least significant byte in the PC by reading the register PCL, but it is not possible to know the value of the 5 most significant bits of PC because PCLATH is not updated from the PC.

When the program is being executed, the PC is incremented by 1 unit with every instruction, with the exception of those instructions that modify the content of the PC. Instructions such as goto (unconditional jump), call (subroutine call), return, retfie, retlw, and other instructions whose destination is the PCL register, modify the content of the PC register. In these cases, the relationship between the PC and the registers PCLATH and PCL is as follows:

- When executing an instruction whose destination is the PCL register (indirect jump to the address pointed by PCLATH and PCL), the 8 least significant bits of the PC are loaded with the result of the instructions, while the 5 most significant bits of the PC are loaded with the 5 least significant bits of the PCLATH register as shown in Figure 3.8a.

- When executing a goto or call instruction, the 11 least significant bits of the PC come from the instruction, while the 2 most significant bits are loaded from bits 4 and 3 in PCLATH as shown in figure 3.8b. These 2 most significant bits indicate the page number.

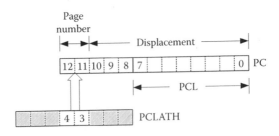

FIGURE 3.7
Program counter (PC) and other components in a program memory address: page number (2 bits) and displacement (11 bits). The page number is loaded from bits PCLATH<4:3>. The 8 least significant bits in the PC make up the register PCL. PCLATH and PCL are special function registers located in the microcontroller's data memory.

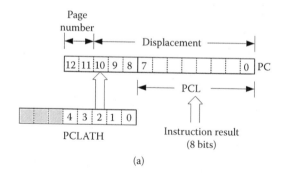

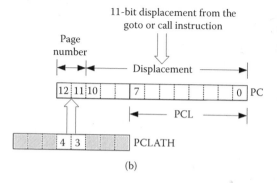

FIGURE 3.8

Program counter (PC) and special function registers PCLATH and PCL. Figure shows the results when (a) executing an instruction whose destination is the PCL or (b) executing a goto or call instruction.

- When executing a return instruction from a subroutine, the 13 bits in the PC are loaded from the top of the stack. PCLATH is not used in this case.

If the program is longer than 1 page, it is necessary to be very careful with the jumps between pages and calls to subroutines stored in a different page. It is then necessary to update bits PCLATH <4:3> correctly before changing pages.

3.2.1.2 Reading and Writing the Program Memory

The instruction set in medium-end PIC microcontrollers does not have a specific instruction to read its program memory. This means that program memory can only be used to store instructions but not data. However, in some applications it can be very useful to have data permanently stored in this memory, such as tables and text created using ASCII characters. For this reason, some medium-end PIC microcontrollers allow indirect read of program memory. This is done by using special function registers in the data memory. Moreover, some PIC microcontrollers using flash memory

TABLE 3.3

Special Function Registers (SFRs) Used in Reading or Reading/Writing into Program Memory in PICs That Allow It

SFR in Devices That Only Allow Reading Program Memory		SFR in Devices That Allow Reading and Writing Program Memory	
PMADRH	Address (13 bits): PMADRH:PMADR	EEADRH	Address (13 bits): EEADRH:EEADR
PMADR		EEADR	
PMDATH	Data (14 bits): PMDATH:PMDATA	EEDATH	Data (14 bits): EEDATH:EEDATA
PMDATA		EEDATA	
PMCON1	Control	EECON1	Control
		EECON2	

also allow writing of data in the program memory using a similar indirect approach. Examples of these devices are PIC16C781 and PIC16C782, which allow the indirect reading of program memory, and the PIC16F87x family, which allows writing and reading of flash program memory.

Table 3.3 shows the special function registers used in reading or reading/writing program memory. The registers PMDATH:PMDATA or EEDATH:EEDATA are used to read or write the 14-bit data. The registers PMADRH:PMADR or EEADRH:EEADR are used for the 13-bit data address, and the registers PMCON1 or EECON1 and EECON2 are the control registers.

The procedure for reading program memory is to:

1. Write the address of the cell to read in the special function registers PMADRH:PMADR or EEADRH:EEADR depending on the type of device used. In PICs that allow writing and reading of program memory, it is necessary to set the bit EEPGD to 1 in register EECON1. This indicates that program memory will be accessed instead of the EEPROM data memory.

2. Set the bit RD in the register PCON1 (or EECON1) to 1. This initiates the reading process. Reading takes two instruction cycles, in which, although PC is increased in two cycles, there is no instruction fetching. For this reason, it is necessary to put two additional instructions after setting bit RD to 1, although these two instructions will not be executed. The best option is to place the nonoperation instruction (nop).

3. Once the reading operation is complete, the bit RD is set to 0 automatically. Also, the bit EEIF in the register PIR2 is set to 1, thus indicating the end of the reading operation.

4. The registers PMDATH:PMDATA or EEDATH:EEDATA contain the data in the original cell that was read.

Example 3.1

The following is a segment of program code that shows how to read program memory in a PIC16F873 microcontroller. This specific microcontroller allows writing and reading into its program memory.

```
bsf     STATUS, RP1     ; Select bank 2.
bcf     STATUS, RP0     ;
movf    ADDR_H, W       ; Write address in
movwf   EEADRH          ; EEADRH:EEADR.
movf    ADDR_L, W       ;
movwf   EEADR           ;
bsf     STATUS, RP0     ; Select bank 3.
bsf     EECON1, EEPGD   ; Select program memory.
bsf     EECON1, RD      ; Start reading operation.
nop                     ; Sequence of 2 nop required.
nop
bcf     STATUS, RP0     ; Select bank 2 as
                        ; data is ready in EEDATH:EEDATA.
movf    EEDATH, W       ; Read data.
movwf   DATA_H          ;
movf    EEDATA, W       ;
movwf   DATA_L          ;
```

The program takes the address of the program memory cell from registers ADDR_H:ADDR_L and places it in EEADRH:EEADR using the W register as an intermediate step. Reading starts by setting to 1 the bit RD in EECON1. Because the reading process takes two instruction cycles, it is necessary to use two nop operations. Finally, the data being read is taken from the special function registers EEDATH:EEDATA and is placed in registers DATA_H:DATA_L.

It is necessary to select the appropriate bank before accessing the special function registers. This selection is done with bits RP1, RP0 from the SFT STATUS.

In read-only devices it is necessary to use the special function registers PM and remove the instruction bsf EECON1, EEPGD.

The following is the procedure used to write data into the flash program memory:

1. Write the address of the cell in the special function register EEADRH:EEADR. Write the data to be written in EEDATH:EEDATA.
2. Set the bit EEPGD in the EECON1 register to 1 to indicate that the writing will occur in the program memory instead of in the EEPROM data memory.
3. Set the bit WREN in the special function register EECON1 to 1 to enable writing into the program memory.
4. Disable all interrupts.
5. Write 55h to EECON2.

6. Write AAh to EECON2.

7. Set the bit WR in ECON1 to 1. This will start the writing process that will occur in the next two instruction cycles.

8. Write two nonoperation instructions (nop) in the program.

9. Set the bit WREN in SFR EECON1 to 0 to disable writing in the flash memory. This prevents accidental writing in the program memory.

10. At the end of the writing cycle, the bit WR in EECON1 is automatically set to 0. The bit EEIF in SFR PIR2 is set to 1 indicating the end of writing.

11. Enable interrupts.

Steps 5 to 7 are a set of five instructions that need to be executed without being interrupted. Step 10 is a safety measure to avoid accidentally writing undesired data in the program memory.

Example 3.2

The following is a segment of program code that shows the recommended procedure to write in the flash memory for a PIC16F873 microcontroller.

```
bsf     STATUS, RP1     ; Select bank 2.
bcf     STATUS, RP0     ;
movf    ADDR_H, W       ; Write address in
movwf   EEADRH          ; EEADRH:EEADR.
movf    ADDR_L, W       ;
movwf   EEADR           ;
movf    DATA_H, W       ; Write data in
movwf   EEDATH          ; EEDATH:EEDATA.
movf    DATA_L, W       ;
movwf   EEDATA          ;
bsf     STATUS, RP0     ; Select bank 3.
bsf     EECON1, EEPGD   ; Select program memory and
bsf     EECON1, WREN    ; Enable writing into FLASH memory.
bcf     INTCON, GIE     ; Disable all interrupts.
movlw   55h             ; Required sequence.
movwf   EECON2          ;
movlw   AAh             ; Required sequence.
movwf   EECON2          ;
bsf     EECON1, WR      ; Start write operation.
nop                     ; Required sequence while
nop                     ; writing in memory.
bcf     EECON1, WREN    ; Disable memory write.
bsf     INTCON, GIE     ; Enable interrupts (optional).
```

It is necessary to select the appropriate bank before accessing the special function registers. This selection is done with bits RP1 and RP0 from the SFT STATUS. The bit GIE in the SFR INTCON enables or disables the interrupts.

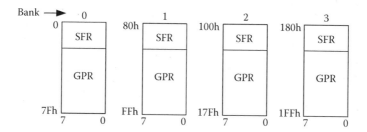

FIGURE 3.9

Data memory paging in medium-end PIC microcontrollers. They can have up to four banks with 128 registers in each one, for a total of 512 registers. Each bank contains special function registers (SFRs) and general purpose registers (GPRs). Each register is 8 bits long.

3.2.2 RAM Data Memory

In PIC microcontrollers, the data memory implemented using RAM technology is organized in 8-bit words. Similar to program memory, data memory is also divided in different pages; for data memory these pages are called *banks*. Each bank can have up to 128 memory cells or registers. The range of addresses for each bank is from 00h to 7Fh.

All medium-end PICs have at least two memory banks. This gives them up to 256 registers with absolute addresses ranging from 00h to FFh. Some devices have up to four memory banks, thus giving a total of 512 registers with absolute addresses ranging from 000h to 1FFh as shown in figure 3.9.

Data memory registers can be classified as special function registers (SFRs) or general purpose registers (GPRs). SFRs are used to control the PIC and access its peripheral modules. GPRs make up the data memory available to the user. The size of SFRs and GPRs is dependent on the type of PIC used. Figures 3.10 and 3.11 show the data memory registers for the microcontrollers PIC16F84 and PIC16F873, respectively. The PIC16F84 has 14 FSRs and 68 GPRs. The PIC16F873 has 50 SFRs and 192 GPRs. It is important to note that to ease the writing of the program, it is possible to access most of the special function registers from any memory bank.

3.2.2.1 Addressing Data Memory

The address of a memory cell needs 9 bits. Due to the paging system used, bits 8 and 9 indicate the bank number and bits 6 to 0 indicate the address within that bank (displacement), as shown in Figure 3.12. The two bits that identify the bank come from the STATUS special function register. The displacement can be located either in the instruction (direct addressing) or in the special function register FSR (indirect addressing).

Either direct or indirect addressing can access all registers in the data memory as shown in Figure 3.13. When using direct addressing, the bank

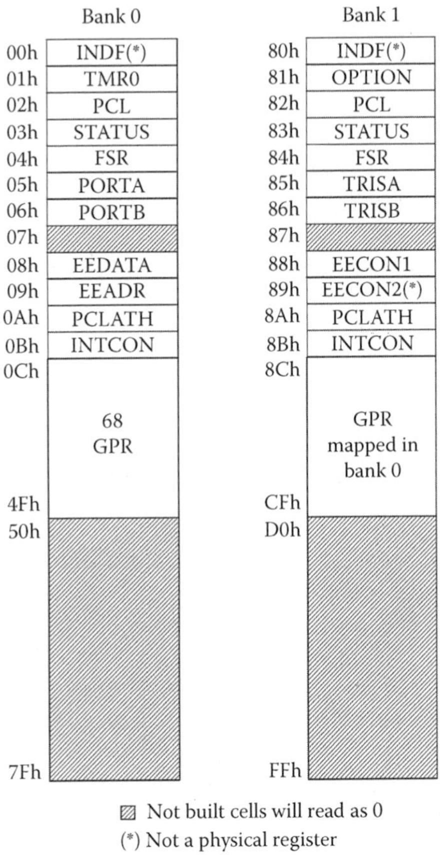

Bank 0		Bank 1	
00h	INDF(*)	80h	INDF(*)
01h	TMR0	81h	OPTION
02h	PCL	82h	PCL
03h	STATUS	83h	STATUS
04h	FSR	84h	FSR
05h	PORTA	85h	TRISA
06h	PORTB	86h	TRISB
07h		87h	
08h	EEDATA	88h	EECON1
09h	EEADR	89h	EECON2(*)
0Ah	PCLATH	8Ah	PCLATH
0Bh	INTCON	8Bh	INTCON
0Ch		8Ch	
	68 GPR		GPR mapped in bank 0
4Fh		CFh	
50h		D0h	
7Fh		FFh	

▨ Not built cells will read as 0
(*) Not a physical register

FIGURE 3.10
Data memory registers in the PIC16F84 microcontroller. These are organized in two banks with 68 general purpose registers (GPRs) that can be accessed from any bank, and 14 special function registers (SFRs). Some SFRs such as STATUS, PCLATH, PCL, FSR, and INTCON can be accessible from any bank. The first cell in each bank (INDF) is used for a data indirect address and it is not a real register in the microcontroller.

is selected with RP1 and RP0 that are bits 6 and 5 in the special function register STATUS. The instruction handles the 7-bit displacement that can range from 00h to 7Fh. For microcontrollers with only two memory banks, the bit RP1 can be ignored.

When using indirect addressing, the 8 least significant bits for the address come from the file select register (FSR). The ninth bit is IRP (Indirect Register Pointer bit), which is bit 7 in the STATUS register. In this case, the 8-bit displacement is in the FSR register. In microcontrollers with only two memory banks, IRP needs to be kept at 0. In microcontrollers with four memory banks, IRP = 0 selects banks 0 and 1; IRP = 1 selects banks 2 and 3. For indirect addressing the data memory can be seen as

Bank 0		Bank 1		Bank 2		Bank 3	
INDF(*)	80h	INDF(*)	100h	INDF(*)	180h	INDF(*)	
TMR0	81h	OPTION	101h	TMR0	181h	OPTION	
PCL	82h	PCL	102h	PCL	182h	PCL	
STATUS	83h	STATUS	103h	STATUS	183h	STATUS	
FSR	84h	FSR	104h	FSR	184h	FSR	
PORTA	85h	TRISA	105h	/////	185h	/////	
PORTB	86h	TRISB	106h	PORTB	186h	TRISB	
PORTC	87h	TRISC	107h	/////	187h	/////	
PORTD(¹)	88h	TRISD(¹)	108h	/////	188h	/////	
PORTE(¹)	89h	TRISE(¹)	109h		189h	/////	
PCLATH	8Ah	PCLATH	10Ah	PCLATH	18Ah	PCLATH	
INTCON	8Bh	INTCON	10Bh	INTCON	18Bh	INTCON	
PIR1	8Ch	PIE1	10Ch	EEDATA	18Ch	EECON1	
PIR2	8Dh	PIE2	10Dh	EEADR	18Dh	EECON2	
TMR1L	8Eh	PCON	10Eh	EEDATH	18Eh	(²)	
TMR1H	8Fh	/////	10Fh	EEADRH	18Fh	(²)	
T1CON	90h	/////	110h	/////	190h	/////	
TMR2	91h	SSPCON2		/////		/////	
T2CON	92h	PR2		/////		/////	
SSPBUF	93h	SSPADD		/////		/////	
SSPCON	94h	SSPSTAT		/////		/////	
CCPR1L	95h	/////		/////		/////	
CCPR1H	96h	/////		/////		/////	
CCP1CON	97h	/////		/////		/////	
RCSTA	98h	TXSTA		/////		/////	
TXREG	99h	SPBRG		/////		/////	
RCREG	9Ah	/////		/////		/////	
CCPR2L	9Bh	/////		/////		/////	
CCPR2H	9Ch	/////		/////		/////	
CCP2CON	9Dh	/////		/////		/////	
ADRESH	9Eh	ADRESL		/////		/////	
ADCON0	9Fh	ADCON1	11Fh	/////	19Fh	/////	
	A0h		120h		1A0h		
96 GPR		96 GPR		GPR mapped in bank 0		GPR mapped in bank 1	
	FFh		17Fh		1FFh		

▨ Not built cell will read as 0   (¹) Not used in PIC16F873
(*) Not a physical register   (²) Reserved

FIGURE 3.11
Data memory registers for the PIC16F873. These are organized in four banks having 192 general purpose registers and 50 special function registers.

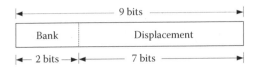

FIGURE 3.12
Data memory address components. Nine bits are required: the 2 most significant bits indicate the bank number and the 7 least significant bits indicate the displacement or address within the bank.

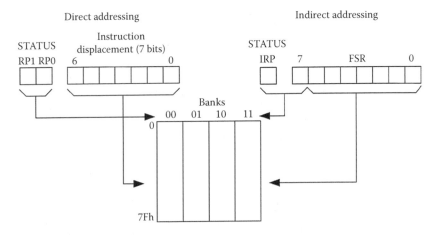

FIGURE 3.13
Data memory addressing. All memory registers can be accessed using direct or indirect addressing.

paged into two banks, each one with 256 addresses, instead of the four banks with 128 addresses as used in direct addressing.

3.2.2.2 Special Function Registers (SFRs)

SFRs are registers located in the data memory with specific information or control functions for the microcontroller or its peripherals. The SFRs associated with the core of the microcontrollers are common among the different devices. On the other hand, the SFRs associated with peripherals are strongly dependent on those peripherals.

Table 3.4 shows the SFRs associated with the different functions and peripherals in medium-end PIC microcontrollers. The STATUS, PCLATH, PCL, FSR, OPTION, INTCON, PORTA, PORTB, TRISA, TRISB, and TMR0 registers are common to most of them.

3.2.2.2.1 The STATUS Register

The STATUS register contains the bits associated to arithmetic operations as well as the bits for selecting the banks of memory. It also contains two

TABLE 3.4

Special Function Registers (SFRs) Associated with the Microcontroller or Its Peripheral

Function or Device	SFR
Select memory bank. Indicators related to arithmetic and logic operations. Watchdog timer overflow. Low-power indication.	STATUS
Prescaler value. Edge for clock pulses. Edges for external interrupt requests. Internal pull-up for port B.	OPTION
Memory parity error indicators. Type of reset. Low-power.	PCON
Program counter	PCLATH, PCL
Indirect addressing	FSR
Interrupts	INTCON
	PIR1, PIE1
	PIR2, PIE2
Parallel ports	PORTA, TRISA
	PORTB, TRISB
	PORTC, TRISC
	PORTD, TRISD
	PORTE, TRISE
Timer0	TMR0, OPTION, INTCON
Timer1	TMR1H, TMR1L, T1CON, PIR1
Timer2	TMR2, PR2, T2CON, PIR1
Modules CCPx (x = 1, 2, 3)	CCPxH, CCPRxL,
	CCPxCON
Serial port USART or SCI	TXREG, TXSTA, RCREG, RCSTA
Synchronous serial port SSP	SSPSTAT, SSPCON, SSPBUF, SSPADD
A/D converter	ADRESH, ADRESL, ADCON0, ADCON1
EEPROM data memory and program flash memory	EEADRH, EEADR, EEDATH, EEDATA, EECON1, EECON2

bits that indicate the state of the watchdog timer and the low-power mode. It is possible to access the STATUS register from any memory bank. This register is present in all the memory banks because it is used to select the banks, and therefore needs to be accessible from any bank. Figure 3.14 shows the bits in the STATUS register.

IRP: This bit selects the memory bank in indirect addressing. IRP = 0 selects banks 0 and 1. IRP = 1 selects banks 2 and 3.

STATUS
(address 03h in any bank)

7	6	5	4	3	2	1	0
IRP	RP1	RP0	TO#	PD#	Z	DC	C
R/W-0	R/W-0	R/W-0	R-1	R-1	R/W-x	R/W-x	R/W-x

FIGURE 3.14
STATUS register. This register can be accessed at address 03h from any data memory bank. This figure shows the values of the bits in this register after a reset. It also shows which bits are read/write (R/W) and which bits are read-only (R).

RP1, RP0: These bits are used to select the banks in direct memory addressing: 00 = bank 0; 01 = bank 1; 10 = bank 2; 11 = bank 3.

TO#: This bit indicates the state of the WDT. When TO = 0 it means that the WDT had overflow. TO is set to 1 with a power-on reset as well as with the instructions clrwdt and sleep.

PD#: Indication of low-power mode or power-down. This bit is set to 0 when the microcontroller enters a low-consumption mode as a result of the instruction sleep. It is set to 1 with the instruction clrwdt as well as with a power-on reset.

Z: Zero indicator. Z = 1 indicates that the result of an arithmetic or logic operation was zero. Otherwise, Z = 0.

DC: Digit carry bit. DC = 1 when there is a carry over between bits 3 and 4 in binary addition. Otherwise DC = 0. In a subtraction operation DC = 0 if there is borrowing between bits 4 and 3 and DC = 1 when there is no borrowing.

C: Carry bit. C = 1 when there is carry over between bits 7 and 8 in binary addition. Otherwise DC = 1. In a subtraction operation, C = 0 when there is borrowing between bits 8 and 7 and C = 1 when there is no borrowing.

As with any other register, the STATUS register can be the destination for any instruction. In this case it is necessary to keep in mind that the resulting value in this register may be different from the value it intended to write because the bits TO# and PD# are read-only bits and therefore can not be written by the program. The bits Z, DC, and C will be set to the appropriate values according to the instruction logic instead of the values that the instruction is supposed to write.

For example, the instruction clrf STATUS will not put 00h in STATUS but will leave TO# and PD# unmodified. Also, bits DC, C, and Z will remain unchanged at 1 according to the logic of this instruction. Therefore, the resulting value in STATUS will be 000uu1uu (with u meaning "unchanged" bit).

OPTION
(address 01h in banks 1 and 3)

7	6	5	4	3	2	1	0
RBPU#	INTEDG	T0CS	T0SE	PSA	PS2	PS1	PS0
R/W-1	R/W-1	R/W-1	R/W-1	R/W-1	R/W-1	R/W-1	R/W-1

FIGURE 3.15

OPTION register. The figure shows the values of the bits in this register after a reset. All the bits are read/write bits (R/W).

To modify the bits of the STATUS register it is recommended to use instructions that do not change bits Z, DC, or C, such as the instructions bcf, bsf, swapf, and movwf.

3.2.2.2.2 The OPTION Register

The OPTION register (also known as OPTION_REG) contains the bits to control additional functions related to enabling the internal pull-up in port B, the edge for recognizing external interrupts, the source of pulses for Timer0, the selection of prescaler for Timer0, and the watchdog timer. Figure 3.15 shows the OPTION register.

RBPU#: Internal pull-up in port B. RBPU# = 1 disables the internal pull-up in port B. With RBPU# = 0, the pull-ups for each bit in port B can be enabled individually.

INTEDG: Edge for external interrupt. With ENTEDG = 1, the external interrupt is produced with the rising edge of the signal; when this bit set to 0, it happens with the falling edge of the signal.

T0CS: Selection of clock source for Timer0. This bit selects the source for the pulses in Timer0. T0CS = 1 selects the pulses in pin T0CK1 as the source of pulses. T0CS = 0 selects the internal clock (divided by 4).

T0SE: Selects edge in Timer0 clock. T0SE = 1 increments Timer0 with the raising edge and T0SE = 0 increments Timer0 with the falling edge.

PSA: Prescaler assignment. PSA = 1 assigns the prescaler to WDT. PSA = 0 assigns the prescaler to Timer0.

PS2, PS1, PS0: Bits to select the prescaler:

PS2	PS1	PS0	Division with Prescaler Assigned to Timer0	Division with Prescaler Assigned to WDT
0	0	0	2	1
0	0	1	4	2
0	1	0	8	4
0	1	1	16	8
1	0	0	32	16
1	0	1	64	32
1	1	0	128	64
1	1	1	256	128

3.2.3 EEPROM Data Memory

Most of the medium-end PICs that use flash memory also include up to 256 bytes of non-volatile memory for data or EEPROM data. This memory is physically separated from the RAM data memory. This memory is accessed by special function registers, similar to those used to read and write in the program flash memory. Table 3.5 shows these registers and their function in handling EEPROM data memory.

The procedure to read data stored in the EEPROM memory is as follows:

1. Write the address of the memory cell to be read in the EEADR.

2. Set the bit EEPGD in the register EECON1 to 0. This indicates the access to EEPROM data memory instead of the program memory.

3. Set the bit RD in the register EECON1 to 1. This starts the reading process. Data will be available in the register EEDATA in the next instruction cycle.

4. Once the reading operation has finished, bit RD is automatically set to 0. Also, bit EEIF in register PIR2 is set to 1, thus indicating the end of the reading operation.

TABLE 3.5

Special Function Registers Used for EEPROM Data Memory Operations

Register	Function
EEADR	Contains data address (8 bits)
EEDATA	Contains the data (8 bits)
EECON1	Control
EECON2	

Example 3.3

The following is a segment of program code that shows the recommended procedure to read a cell in the EEPROM data memory in a PIC16F873 microcontroller.

```
bsf     STATUS, RP1     ; Select bank 2.
bcf     STATUS, RP0     ;
movf    ADDR, W         ; EEPROM address is in ADDR.
movwf   EEADR           ; and is placed in EEADR.
bsf     STATUS, RP0     ; Select bank 3.
bcf     EECON1, EEPGD   ; Select data memory.
bsf     EECON1, RD      ; Start reading operation.
bcf     STATUS, RP0     ; Select bank 1 as data is ready
movf    EEDATA, W       ; in EEDATA. Move it to W.
```

The program takes the address of the cell in the EEPROM memory that is in the register ADDR and places it in EEADR using the W register as an intermediate step. Reading starts by setting to 1 the bit RD in EECON1. The data being read is taken from the special function register EEDATA and is placed in the W register. Before accessing any register, it is necessary to select the appropriate bank by means of the bits RP1 and RP0 in the SFR STATUS.

The procedure to write data in a cell in the EEPROM data memory is as follows:

1. Write the address of the memory cell in the special function register EEADRH and the data to be written in EEDATA.
2. Set the bit EEPGD in the EECON1 register to 0. This indicates that the EEPROM data memory will be accessed instead of the program flash memory.
3. Set the bit WREN in SFR EECON1 to 1 in order to enable writing in the EEPROM data memory.
4. Disable all interrupts.
5. Write 55h in EECON2.
6. Write AAh in EECON2
7. Set the bit WR in SFR ECON1 to 1. This begins the writing process. This process takes place during the following two instruction cycles.
8. Set the bit WREN in SFR EECON1 to 0 to disable accidental writing in the EEPROM data memory.
9. Once the writing process is finished, the bit WR in EECON1 is automatically set to 0. The bit EEIF in SFR PIR2 is set to 1. This indicates that the reading process has finished.
10. Enable interrupts.

Steps 5 to 7 are a set of five instructions that need to be executed without being interrupted. Step 10 is a safety measure to avoid writing undesired data in the program memory accidentally.

Example 3.4

The following is a segment of program code that shows the recommended procedure to write in the EEPROM memory for a PIC16F873 microcontroller.

```
            bsf       STATUS, RP1     ; Select bank 2.
            bcf       STATUS, RP0     ;
            movf      ADDR, W         ; Write address in
            movwf     EEADR           ; EEADR.
            movf      DATA, W         ; Write data in
            movwf     EEDATA          ; EEDATA.
            bsf       STATUS, RP0     ; Select bank 3.
            bcf       EECON1, EEPGD   ; Select EEPROM data memory.
            bsf       EECON1, WREN    ; Enable memory write.
            bcf       INTCON, GIE     ; Disable interrupts.
            movlw     55h             ; Required sequence.
            movwf     EECON2
            movlw     AAh             ; Required sequence.
            movwf     EECON2
            bsf       EECON1, WR      ; Start write operation.
            bcf       EECON1, WREN    ; Disable memory writing.
wait_time:
            btfsc     EECON1, WR      ; ¿WR = 0 ?if yes, continue.
            goto      wait_time       ; if not, continue waiting.
            bsf       INTCON, GIE     ; Enable interrupts (optional)
```

It is necessary to select the appropriate bank before accessing the special function registers. This selection is done with bits RP1 and RP0 from the SFT STATUS. The bit GIE in the SFR INTCON enables or disables the interrupts.

4

Instruction Set and Assembler Language Programming

This chapter describes the processes for designing and writing programs in assembler language for PIC microcontrollers. The chapter starts by describing basic concepts such as machine language, assembler language, and source and object codes. It continues by explaining the instruction set in medium-end PIC microcontrollers, using several examples to show the reader how these instructions are used. Finally, this chapter describes available resources for developing programs in assembler language using a personal computer.

4.1 Basic Concepts

4.1.1 Machine Code and Assembler Language

All microprocessors and microcontrollers execute their instructions by using their own machine language. *Machine language* is made of the binary codes that describe the instructions for the device to execute. It is a language made of 1s and 0s. For example, because each instruction in medium-end PIC microcontrollers has 14 bits, a program in machine language for these microcontrollers consists of 14-bit words.

Obviously, the development of programs in machine language can be extremely difficult. To ease the task of writing code at this "low level," assembler language was developed. *Assembler language* consists of mnemonic symbols that represent the corresponding machine language bits.

Example 4.1

The instruction "set W register to 0" is represented in machine language as:

```
00 0001 0xxx xxxx
```

with x being 0 or 1 indistinctly.

The same operation in assembler language is represented using the mnemonic clrw (clear W register), thus making assembler language much easier to work with compared to machine language.

Example 4.2

The instruction "store the value K in register W" with K being an 8-bit binary number is represented in machine language as:

```
11 00xx kkkk kkkk
```

where k are the binary digits of K and x could be indistinctly 0 or 1.

In assembler language the instruction "store an 8 bit data in the W register" is represented by the mnemonic movlw (move literal to W). Therefore, the complete instruction becomes:

```
movlw K
```

This instruction is much easier to work with than machine language.

Both, machine and assembler languages are highly dependent on each family of microprocessors or microcontrollers. Each type of microcontroller has its own assembler language, normally different from the assembler language in another type of microcontroller. Medium-end PICs have an assembler language with 35 instructions.

Microcontrollers cannot directly execute a program written in assembler language. It is necessary to "translate" this language into machine language. This process is called *assembly* and it is done by a program called an *assembler*, although some additional programs may also intervene. The original program written in assembler language is called *source code* and the code after the assembling process is called *object code*. This assembling process is normally carried out by a personal computer. There are several methods to obtain machine code from the source code written in assembler language. The way the assembler operates depends on how the source code was written, and in particular it is dependent on whether the source code contains the real addresses for storing instructions and data. If the source code specifies the real addresses of data and instructions (direct addresses), then the assembler generates *absolute object code*; otherwise, the assembler generates *relative object code* also known as *relocatable code*. Figure 4.1 illustrates these two assembly modes.

When the source code contains all the information needed to translate the program into machine language, that is, if it contains the real data and instruction addresses, as well as the addresses they will occupy in memory, the assembler can directly generate the object code. This method of producing programs that can be directly coded by the assembler is only

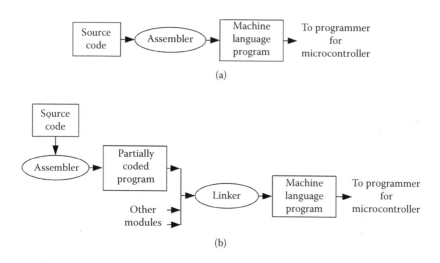

FIGURE 4.1
Process for translating a program written in assembler into machine language. In (a) the assembler directly generates the object program in machine language. In (b) the assembler partially codifies the source code program, and the linker generates the object code in machine language through the different modules.

useful when the program is short and simple. Otherwise this way of programming is not flexible or practical.

Larger projects require a modular approach. This is done by developing programs for each separate module and then linking them all together to create the object code. When writing the programs for each separate module, it is useful not to indicate the direct memory addresses to store instructions or data. With this, the assembler can only create a partial or indirect code because it does not have the real memory addresses where information will be stored. In this case, an additional program called a *linker* is needed. The linker takes the different modules that have been created, sets the real addresses in memory for instructions and data by linking all the modules, and finally creates the code object program in machine language.

The use of a linker introduces additional flexibility and increase of power in the process of writing programs for microcontrollers; it also requires writing them in a manner different from how they would be written when creating absolute object code. The linker also provides the advantage of allowing the use of modules written in different languages such as C and assembler. Some of these modules may constitute a program library. For example, one specific module can contain all the subroutines used for floating point arithmetic operations. When linking this module with the rest of the modules, the linker will only take from the whole library the subroutines that are required for that program to create the object code. All this contributes to increase the flexibility and power of the modular approach for programming microcontrollers.

4.1.2 Structure of Instructions

The instructions in any microprocessor or microcontroller have two components: the *operation code* and the *operands*. The operation code contains the order for the microcontroller to execute the operation described by the instruction. The operands are the data needed in carrying out the instruction. Operands can be addresses or data. The operands for medium-range PIC microcontrollers can be:

- A 7-bit address in the data memory
- An 11-bit address in the program memory
- An 8-bit data
- A 3-bit address of a bit register in the data memory
- The 1-bit indication of whether the result of the instruction will be placed in the working (W) register or in the data memory

In general, some instructions may not require operands, whereas other instructions may require more than one operand. Medium-end PIC microcontrollers have instructions with no operands and instructions with one or two operands. Figure 4.2 illustrates different possible cases for instructions using one or two operands. These can be:

- Instructions that carry out operations using registers in the data memory have two operands. One operand is the 7-bit register address. The second operand is the bit that indicates whether the destination of the result will be stored in the W register or in the register indicated in the original instruction.
- Instructions that include an 8-bit data in the instruction only need this data as an operand.
- Instructions that include the 11-bit program memory address have this address as their only operand.
- Instructions that carry out operations with the bits in the data memory need two operands. One operand is the address of the bit within the register. The second operand is the 7-bit address for the register in the data memory.

All the instructions in medium-end PIC microcontrollers are 14 bits long. This is also the length of the cells in the program memory and there-

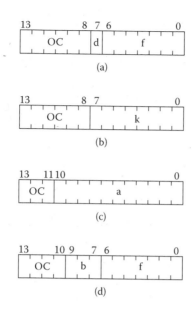

FIGURE 4.2

Format of instructions in medium-end PIC microcontrollers. Instructions have operation codes from 3 to 6 bits with up to two operands. (a) Format for instructions that work with data memory registers. (b) Instructions that work with 8-bit immediate data. (c) Instructions containing program memory addresses. (d) Instructions for manipulating a specific bit in a data memory register. OC, operation code; f, 7-bit address of a register in data memory; k, 8-bit data; a, 11-bit address; b, number of bit (0 to 7); d, destination of result: if d = 0, destination is W and if d = 1 destination is f.

fore each instruction can be stored in a single memory cell. This structure is commonly used in RISC microprocessors and microcontrollers.

4.1.3 Data Addressing Modes

Data addressing modes refers to the different ways in which data can be addressed within an instruction. When writing an instruction, data can be referred to in two ways: in the instruction itself (as an operand) or through its specific address in data memory.

The modes for data addressing are strongly dependent on the architecture of the device. In the simplest form, data addressing can be either in the instruction or in a register. This is the approach used in medium-end PIC microcontrollers. The register that stores data addresses is generically called the data address register. There are three basic modes to refer data: with data in the instruction, with the data address in the instruction, and with the address for the data in the data address register. Therefore, there are three possible modes to addressing data: *immediate addressing, direct addressing,* and *indirect addressing.*

In immediate addressing, the data is part of the instruction. In this case, the instruction operand is the data itself. In direct addressing, the address of the data is part of the instruction. In this case, the instruction operand is the data address. In indirect addressing, the instruction takes the data address from the data address register. In this case, the instruction operand is the address of the data address register.

In medium-end PIC microcontrollers the data address register is its file select register (FSR). All data memory registers in medium-end PIC microcontrollers can be accessed by using direct or indirect addressing. When using these addressing methods it is necessary to remember that memory is organized in banks. Therefore, the first step is to select the bank that contains the register of interest. Banks are selected using bits IRP, RP1, and RP0 in the special function register STATUS as shown in figure 3.13. When using direct addressing, the bank is selected with bits RP1 and RP0. When using indirect addressing, the banks are selected with IRP and the most significant bit in the FSR. The FSR needs to contain the address of the register to be addressed. Data is read or written in the register pointed to by the FSR when the data is read or written in the register INDF (Indirect File).

The following examples illustrate direct and indirect addressing in medium-end PICs.

Example 4.3

Given a PIC16F84, store the data 0x35 in the W register.
Solution: movlw 0x35. This uses immediate addressing as the data is contained in the instruction itself.

Example 4.4

Given a PIC16F873 microcontroller, place the value 0x35 in the register 0x20 of bank 1. Do this task using direct and indirect addressing.
Solution using direct addressing: Bank 1 will be selected first and then the data will be written in the register using an instruction that has the register address (0x20) as an operand:

```
bcf    STATUS, RP1    ; Select bank 1.
bsf    STATUS, RP0
movlw 0x35            ; Place the value 0x35 in W and
movwf 0x20            ; copy it in the register whose address is 0x20
                     ; in the selected bank.
```

Solution using indirect addressing: Bank 1 is selected with IRP = 0 and placed in the FSR the register address (0x20) with the most significant bit in the FSR equal to 1, that is, 0xA0. The data 0x35 is placed in the register pointed to by FSR when the instruction that writes in INDF is executed.

```
bcf    STATUS, IRP    ; Select banks 0 and 1.
movlw 0xA0            ; Place the register address in W and
movwf FSR            ; copy it to FSR.
movlw 0x35           ; Place 0x35 in W and
movwf INDF           ; copy it in the register pointed to by FSR,
                     ; indicated by writing in INDF.
```

4.1.4 The Stack

The stack is a data storage structure using a last in, first out (LIFO) approach: the last data entering the stack is the first data leaving the stack. The stack has a *base* and a *top*. The base of the stack is the oldest data inside, and the top of the stack is the newest data. When the stack contains multiple data, all the storage or retrieve operations are carried over the top of the stack. The depth of the stack is the size of the stack in a given moment. In most microprocessors and microcontrollers, the stack is a part of the data memory, giving the stack a practically unlimited size. These devices have a stack pointer (SP) register that contains the address of the top of the stack. Storing or retrieving data modifies the SP. For example, when storing data in the stack, its SP increases; whereas when retrieving data from the stack, its SP decreases. Figure 4.3 shows the general structure of the stack in a microprocessor or microcontroller.

The stack is used to store instruction addresses and in particular to "remember" the address to return to the main program from a subroutine. When a call or similar instruction calls a subroutine, the value of the program counter (PC) is stored in the stack. The PC is the address to which the program must return once it has finished executing the subroutine. When the subroutine ends with the instruction return or by using

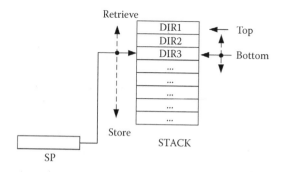

FIGURE 4.3

General structure of the stack in a microprocessor or a microcontroller. The stack is located in a region within the RAM memory. The stack pointer (SP register) points toward the bottom of the pile. This is variable because the stack grows when storing additional data and shrinks when data is retrieved. DIR1, DIR2, and DIR3 are addresses stored in the stack.

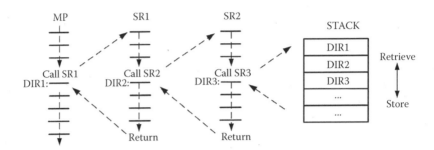

FIGURE 4.4
The stack and subroutine nesting. The main program (MP) calls subroutine SR1, which calls SR2, and so on. Every time the instruction call is executed it stores the return address in the stack. Each return address (DIR1, DIR2, DIR3, …) is equal to the value of the program counter (PC) when the instruction call was executed. The instruction return, which finishes the subroutine, retries the return address and stores it in the PC. The last address stored in the stack is the first address to be retrieved, using the LIFO structure of the stack.

a similar instruction, the instruction takes the value located at the top of the stack and places it in the PC. This causes the program to execute the instruction immediately after the call instruction. The LIFO structure of the stack allows subroutine nesting, that is, the call to a subroutine from another subroutine as shown in Figure 4.4.

In medium-end PIC microcontrollers, the stack has some specific characteristics:

1. The stack is separated from the data memory and program memory.
2. There is no stack pointer register.
3. The stack can only store addresses.
4. The size of the stack is limited, up to eight addresses.

With this, it is possible to represent the stack in medium-end PIC microcontrollers as a set of eight registers, each one of them being 13 bits long. These registers, which store addresses from the program memory, are organized following a LIFO structure. The instructions that manipulate the stack are: call, return, retfie, and retlw. The stack also stores addresses in the event of interrupts.

12		0
~~DIR1~~ DIR9		
~~DIR2~~ DIR10		
~~DIR3~~ DIR11		
DIR4		
DIR5		
DIR6		
DIR7		
DIR8		

FIGURE 4.5
The stack in medium-end PIC microcontrollers is made of eight registers or cells, each one with 13 bits. The stack overflows when more than eight subroutines are nested.

The limited size of this stack allows the nesting of up to eight subroutines. That is, the main program can call a subroutine; this subroutine can call another subroutine and so on up to a total of eight calls. This eight-level stack depth is more than enough in the majority of applications. However, it is the responsibility of the programmer to ensure that the number of nested calls does not exceed the maximum allowable (Figure 4.5). These microcontrollers are not able to indicate stack overflow.

4.2 Instruction Set in Medium-End PIC Microcontrollers

The model for programming medium-end PICs consists of two elements: the W register and the registers in the data memory. The task of the W register is similar to the task of the traditional accumulator in microprocessors. The registers in the data memory can be special function registers (SFRs) or general purpose registers (GPRs) as shown in Figure 4.6.

From the point of view of the programmer, the most important characteristics for a medium-end PIC microcontroller instruction set are:

1. All instructions are 14 bits long.

2. Most of the instructions are executed during a single instruction cycle that lasts four oscillator periods. Only branch instructions and instructions that modify the content of the Program Counter Low (PCL) register need two instruction cycles for their execution.

3. Any register in the microcontroller can be the source or destination for data transfer, arithmetic, or logic operations.

4. Any bit from any registers in the data memory can be individually accessed.

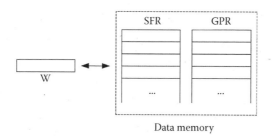

Data memory

FIGURE 4.6

Programming model for medium-end PIC microcontrollers with three types of registers: Work (W) register, special function registers (SFRs), and general purpose registers (GPRs).

TABLE 4.1

Medium-End PIC Microcontroller Instruction Set

Mnemonic		Operation	Affects	Cycles
1. Data Transfer				
movf	f, d	f = > d	Z	1
movwf	f	W = >f	—	1
movlw	k	k = >W	—	1
clrf	f	0 = >f	Z	1
clrw		0 = >W	Z	1
2. Arithmetic and Logic				
addwf	f, d	f + W = > d	C, DC, Z	1
addlw	k	k + W = > W	C, DC, Z	1
subwf	f, d	f – W = > d	C, DC, Z	1
sublw	k	k – W = > W	C, DC, Z	1
incf	f, d	f + 1 = > d	Z	1
decf	f, d	f – 1 = > d	Z	1
andwf	f, d	f and W = > d	Z	1
andlw	k	k and W = > W	Z	1
iorwf	f, d	f or W = > d	Z	1
iorlw	k	k or W = > W	Z	1
xorwf	f, d	f xor W= > d	Z	1
xorlw	k	k xor W = > W	Z	1
rlf	f, d	rotate f left through C = > d	C	1
rrf	f, d	rotate f right through C = >d	C	1
comf	f, d	#f = > d	Z	1
swapf	f, d	$f_L \leftrightarrow f_H$ = > d	—	1
3. Control Transfer				
goto	a	branch to address	—	2
btfsc	f, b	branch if f<b> = 0	—	1(2)
btfss	f, b	branch iff<b> = 1	—	1(2)
incfsz	f, d	f + 1 = > d, branch if 0	—	1(2)
decfsz	f, d	f - 1 = > d, branch if 0	—	1(2)
call	a	call subroutine in address a	—	2
return		subroutine return	—	2
retfie		interrupt return	—	2
retlw	k	return from subroutine with k in W	—	2
4. Bit Manipulation				
bcf	f, b	0 = > f<b>	—	1
bsf	f, b	1 = > f<b>	—	1
5. Other				
nop		no operation	—	1
clrwdt		0 = > WDT	TO#, PD#	1
sleep		go to low power consumption	TO#, PD#	1

Note: W, working register; f, data memory register; k, 8-bit constant; a, 11-bit constant; b, bit; d, destination as follows: If d = 0 destination is W and if d = 1 destination is f. C, DC, Z, TO#, and PD 3 are specific bits within the STATUS register.

5. It is not possible to transfer data from one memory cell to another memory cell using a single instruction. The W register needs to be used as an intermediate step.

6. There are no instructions to store or retrieve data from the stack, such as the PUSH or POP instructions commonly used in microprocessors. The stack only stores addresses of instructions. Furthermore, only subroutine calls or returns can access the stack.

Table 4.1 summarizes the instructions found in medium-end PIC microcontrollers. To better understand these instructions, they have been classified as:

- Data transfer instructions
- Arithmetic and logic instructions
- Control transfer instructions
- Bit manipulation instructions
- Other

4.2.1 Data Transfer Instructions

Table 4.2 shows the data transfer instructions. When using these instructions, any register in the microcontroller can be the source or destination, but it is not possible to transfer data between two registers with a single instruction. It is necessary to use the W register as an intermediate step. If the register is INDF this means that indirect addressing is being used, and the operation will be carried out with the register pointed to by the SFR.

The instructions movf, clrf, and clrw affect the zero flag, that is, bit Z in the STATUS register. The instructions movwf and movlw do not affect any flags. All these instructions are executed in a single instruction cycle.

TABLE 4.2

Data Transfer Instructions

Mnemonic		Operation	Affects	Cycles
movf	f, d	f = >d	Z	1
movwf	f	W = >f	—	1
movlw	k	k = >W	—	1
clrf	f	0 = >f	Z	1
clrw		0 = >W	Z	1

Note: W, working register; f, data memory register; k, 8-bit constant; a, 11-bit constant; d, destination as follows: If d = 0 destination is W and if d = 1 destination is f. Z is a specific bit in the STATUS register.

The instruction movwf copies the content of the W register into the register f in the data memory without altering any flags. On the other hand, the instruction movf f,0 executes the inverse operation, that is, it copies the content of the register f into the W register. In this case, the register f remains unchanged but the zero (Z) flag is affected. The instruction movf f,1 copies the content of the register f into itself but affects the zero flag (Z). This instruction can be used to find out if the value in register f is zero or different than zero.

When programming in assembler language, the parameter d, which indicates the destination of the operation, can be expressed in several forms. For example, the instruction movf f,d can be written in several ways. Let's assume we are working with the register X, this being a general purpose register. If d = 0, then the instructions movf X,0 and movf X,W are equivalent. If d = 1, then the instructions movf X,1, movf X,f, and movf X are also equivalent.

Example 4.5

Let's assume REG1 and REG2 are two general purpose registers in the same data memory bank. Design a program to exchange their contents.

Solution: To interchange the contents of this register we need a third register that will be called TEMP. This register will be used to store data temporarily as the W register is not enough. If REG1, REG2, and TEMP are located in the same data memory bank, we can write:

```
movf    REG1, W
movwf   TEMP
movf    REG2, W
movwf   REG1
movf    TEMP, W
movwf   REG2
```

4.2.2 Arithmetic and Logic Instructions

The arithmetic and logic instructions are shown in Table 4.3 This instruction set includes arithmetic instructions for addition and subtraction as well as incrementing and decrementing. Logic operations include logic negation or complement; operations *or*, *and*, *xor*; bit rotation to the right or left; and swapping nibbles.

Arithmetic instructions affect bits C, DC, and Z in the STATUS register. All logic instructions affect bit Z with the exception of the rotation that affects bit C and nibble swapping that does not affect any of these bits. When using arithmetic and logic operations with two operands, one of them must be placed in the W register while the other can be in the W register or any other register in the data memory. The result of the operation can be placed either in the W register or in any other register. The

TABLE 4.3

Arithmetic and Logic Instructions

Mnemonic		Operation	Affects	Cycles
addwf	f, d	f+ W = > d	C, DC, Z	1
addlw	k	k + W = > W	C, DC, Z	1
subwf	f, d	f – W = > d	C, DC, Z	1
sublw	k	k – W = > W	C, DC, Z	1
incf	f, d	f +1 = >	Z	1
decf	f, d	f – 1 = >	Z	1
andwf	f, d	f and W = > d	Z	1
andlw	k	k and W = > W	Z	1
iorwf	f, d	f or W = > d	Z	1
iorlw	k	k or W = > W	Z	1
xorwf	f, d	f xor W = > d	Z	1
xorlw	k	k xor W = > W	Z	1
rlf	f, d	rotate f left through C = > d	C	1
rrf	f, d	rotate f right through C = > d	C	1
comf	f, d	#f = > d	Z	1
swapf	f, d	$f_L \leftrightarrow f_H$ = > d	—	1

Note: W, working register; f, data memory register; k, 8-bit constant; a, 11-bit constant; d, destination as follows: If d = 0 destination is W and if d = 1 destination is f. C, DC, and Z are specific bits in the STATUS register.

destination of the result is indicated by the parameter d when codifying the instruction. If d = 0 the result of the instruction is placed in W. If d = 1 the result is placed in the register specified by the instruction.

All instructions that work with any data memory register allow direct or indirect addressing. When the register specified by the instruction is INDF, indirect addressing is used. In this case, the operation indicated by the instruction is carried out with the register pointed to by the special function register FSR.

Different than working with similar instructions in other microcontrollers or microprocessors, the carry over bit (bit C in STATUS register) does not intervene directly in the operation indicated by the instruction. However, this bit is affected by the result. This means, for example, that there is not an instruction that adds two registers taking into account the value of the carry over bit (operation f + W + C). Example 4.7 shows how to proceed when it is necessary to perform this addition.

The rotation instructions rlf f,d and rrf f,d rotate the contents of the register indicated by the instruction in one bit to the left or to the right. The rotation instructions are affected by the carry over bit (bit C in STATUS register). This bit operates as an extension of the register f taking the assumed position of bit 8 in this register. If d = 0 the result is placed in W without modifying f. If d = 1 the result is placed in f, thus modifying its initial value.

Example 4.6

Some logic and arithmetic operations using the W register.

```
Increment W:
    addlw 1
Decrement W:
    addlw 0xff
Logic negation (1-complement):
    iorlw 0xff
2-complement for W:
    xorlw 0xff
    addlw 1
Set several bits to 0, for example, bits 3, 2, 1, 0:
    andlw 0xf0
Set several bits to 1, for example, bits 3, 2, 1,0:
    iorlw 0x0f
```

Example 4.7

When adding or subtracting integer numbers it is necessary to take into account the carry over that may have been produced in the preceding bit. One of the bytes will be in W and the other in a generic register that will be called REG. The carry over in the preceding step is bit C in the STATUS register. If we want the result to be placed in W, the operation to carry out is REG + W + C → W. The following segment of code shows how to perform this addition:

```
btfsc STATUS, C    ; Does C = 0? Yes - branch without incrementing W.
addlw 1            ; No - increment W.
addwf REG, W       ; W + REG = > W.
```

4.2.3 Control Transfer Instructions

Table 4.4 shows the control transfer instructions including unconditional branches and conditional branches depending on the state of a bit in a register as well as subroutine calls and returns.

4.2.3.1 Unconditional Branches, Subroutine Calls, and Returns

The instruction goto "a" produces an unconditional branch to the instruction located in the address indicated by the instruction "a." Therefore, the instruction loads the value "a" in the program counter. The instruction call "a" will execute the subroutine located in address "a." The program

TABLE 4.4

Control Transfer Instructions

Mnemonic		Operation	Affects	Cycles
goto	a	branch to address	—	2
btfsc	f, b	branch if f<b> = 0	—	1(2)
btfss	f, b	branch iff<b> = 1	—	1(2)
incfsz	f, d	f + 1 = > d, branch if 0	—	1(2)
decfsz	f, d	f − 1 = > d, branch if 0	—	1(2)
call	a	call subroutine in address a	—	2
return		subroutine return	—	2
retfie		interrupt return	—	2
retlw	k	return from subroutine with k in W	—	2

Note: W, working register; f, data memory register; k, 8-bit constant; a, 11-bit constant; b, bit; d, destination as follows: If d = 0 destination is W and if d = 1 destination is f.

counter (PC) is stored in the stack, and then it places the address "a" in the program counter, thus creating the branch to the subroutine.

These two instructions (goto and call) operate taking into account that memory is organized in pages. The operand for these instructions is an 11-bit word that represents an address within a page. The other two bits in the PC, bits PC<12:11>, come from bits 4 and 3 in the PCLATH register (refer to Section 3.2.1.1 and Figure 3.7). Therefore, if PCLATH is not modified, the branches will occur within the same page. To branch across pages or call subroutines located in different pages it is necessary to previously modify bits 4 and 3 in the PCLATH register with the desired page number. This process can be carried out with the operator HIGH in assembler language as the next two examples illustrate.

Example 4.8

The following segment code illustrates how to branch to an address located in a different page.

```
Prog:
    movlw   HIGH Prog10
    movwf   PCLATH
    goto    Prog10
Prog10:
    ;
    ; Prog10 can be any address in the program memory.
    ;
```

HIGH is an operator in assembler language that makes bits <15:8> in the address represented by the label Prog10 become the data in the instruction movlw. This makes the destination page number to be placed in bits PCLATH<4:3>. When the instruction goto is executed, the bits PCLATH<4:3> are loaded in bits 12 and 11 in the PC.

Example 4.9

The following program code illustrates how to call a subroutine that is located in a page number different from the page number that calls the subroutine.

```
        movlw   HIGH Subroutine    ; Bits <15:8> in the address where
                                   ; the subroutine begins
        movwf   PCLATH             ; are loaded in PCLATH.
        call    Subroutine         ; Call subroutine.
        ;
        ;
    Subroutine:
        ; Beginning of subroutine that can be in any address
        ; in program memory.
        ;
```

The page number in the destination is loaded into bits PCLATH<4:3>. When the call instruction is executed, the bits PCLATH<4:3> are loaded in bits 12 and 11 in the PC.

The instruction goto produces a *direct* unconditional branch because the target address is located in the instruction. This is not the only type of unconditional branch possible in medium-end PIC microcontrollers. Any instruction that modifies the special function register PCL produces an unconditional branch. In this case, it will be an *indirect* unconditional branch because the target address is located in a register instead of in the instruction itself. Because the PCL register is 8 bits long, the branch can be up to 256 addresses. In order to branch farther away, it is necessary to load the PCLATH register correctly. The following example shows an indirect unconditional branch like this.

Example 4.10

Program an unconditional branch to the address noted by label "Prog20" without using the instruction goto. The operators HIGH and LOW in assembly language can be used in this case. The following code segment illustrates this procedure:

```
        movlw HIGH Prog20
        movwf PCLATH
        movlw LOW Prog20
        movwf PCL
    Prog20:
        ;
        ; Prog20 can be any address in the program memory.
        ;
```

HIGH and LOW are assembler language operators. HIGH allows bits <15:8> in the address represented by the label Prog20 to be used as the operand in the instruction movlw. This makes the high part of the branching address to be placed first in the W register and then in PCLATH. The operator LOW makes the bits <7:0> in the address represented by label Prog20 to be used as the operand in the instruction movlw. This makes the low part of the branching address to be placed first in the W register and then in PCL. When PCL is modified, the contents of PCLATH and PCL move to the program counter, producing a branch to Prog20. The instruction movwf PCL needs two machine cycles because it modifies the value of the program counter.

It is also possible to create a branch relative to the value of PCL. PCL can store a base address, to which the instruction addwf PCL,f can add a value thus producing the target address. The following example illustrates the use of relative branches when handling tables stored in the program memory.

Example 4.11

The subroutine Table contains a series of ASCII characters. The position of one character within the table, relative to its beginning, is stored in a general purpose register called INDEX. We want to place in W the character pointed to by INDEX. The following program code shows how to perform this operation by using an indirect branch through PCL.

```
;
; Main program:
;
movlw    HIGH Table      ;Bits <15:8> from the address where the
                         ;table starts
movwf    PCLATH          ;are loaded inPCLATH.
movf INDEX, W            ;Register INDEX points towards the
                         ;interior of the table.
call Table               ;Call to subroutine table.
                         ;W has the value pointed to by INDEX.
;
;Subroutine Table. It can be in any page in the program memory
;as long as it does not exceed
;256 words and it is bounded by 256 addresses. That is, any of
;its instructions can be located
;by only changing the PCL without having to alter the high part
;of PC.
;Inputs: in W the position of the ASCII character.
;Outputs: in W the requested ASCII character.
;
Table:
    addwf    PCL, f
    retlw    'E'
    retlw    'x'
    retlw    'a'
    retlw    'm'
    retlw    'p'
    retlw    'l'
    retlw    'e'
```

First, the example shows the correct way to call a subroutine located in a different page. This is done by loading bits 4 and 3 of PCLATH with the page number. The operator HIGH makes bits <15:8> in the subroutine address become the data in the instruction movlw, placing them in W. W is then copied in PCLATH making the destination page number to be placed in bits PCLATH <4:3>. When the instruction call is executed, the current value of the PC will be stored in the stack and bits PCLATH <4:3> are loaded in bits 12 and 11 in the PC. This creates the branch to the subroutine.

Once in the subroutine, the instruction addwf adds the value in W to the PCL. The result is placed back into the PCL, creating a branch to one of the retlw the instructions. Operand of retlw is the requested ASCII code. For example, if W = 4, the ASCII code for the character "p" will be placed in W.

The subroutine Table may be located in any program memory page as long as it does not exceed 256 words, and is also bounded by 256 addresses. That is, all of its instructions can be accessed with just changing the value of PCL. The instruction addwf PLC,f modifies the value of the PC and therefore needs two instruction cycles. The instruction retlw also needs two instruction cycles. This makes the execution of the subroutine Table last 4 instruction cycles (16 cycles in the main oscillator in the microcontroller).

The instructions return, retfie, and retlw k are placed within a subroutine in order to return to the original program that called the subroutine.

4.2.3.2 Conditional Branches

The following two instructions are the instructions used for conditional branches: btfsc f,b (bit test file and skip if clear) and btfss f,b (bit test file and skip if set). When using these instructions, the program branches if the condition of bit b in file f allows it. The branch in these instructions is very short: if the condition is met, the program will not execute the next instruction. If the condition is not met, the next instruction is executed.

To better understand these instructions, let us consider the following code segment:

```
btfsc              f,b
instruction 1
instruction 2
```

The condition for branching is f<b> zero. If this condition is met, the program branches directly to instruction 2. If it is not met, the program first executes instruction 1. The instruction btfss works in a similar way. The block diagram in Figure 4.7 helps to understand these instructions.

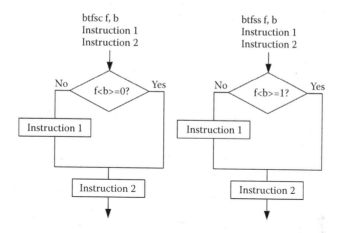

FIGURE 4.7

Conditional jump instructions. The figure shows program segments with the instructions btfsc and btfs as well as their block diagrams.

Example 4.12

Increment W register if carry bit is 1. The following code segment shows the solution to this problem:

```
      btfsc STATUS,C
      addlw 1
Continue: ...
```

The carry bit is bit C in the STATUS register. If the carry over is 1, the condition for branching is not met and therefore the next instruction is executed. This instruction increases the value of the W register in 1 unit. If the carry is 0, the condition is met and therefore the program branches to the label Continue without incrementing W.

The instructions btfsc and btfss are very useful because they can be used to program decisions based on the status of any bit in any register in the microcontroller, either a special function register or a general purpose register. By combining these instructions with the unconditional instruction goto, it is possible to make multiple decisions. The follow example expands on this.

Example 4.13

Figure 4.8 shows the block diagram to program. If the condition f<b> = 1 is met, the program needs to execute action 1. Otherwise, the program needs to execute action 2. After executing any of these instructions, the program has to execute instruction 3.

The following code structure implements the solution to this problem:

```
            btfss f,b
            goto  Action2
Action1:
            ;
            ; Write here instructions for action 1.
            ;
            goto Action3'
Action2:
            ;
            ; Write here instructions for action 2.
            ;
Action3:
            ;
            ; Write here instructions for action 3.
            ;
```

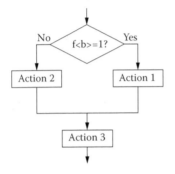

FIGURE 4.8
Block diagram for the algorithm to be programmed in example 4.13.

The instructions incfsz f,d and decfsz f,d combine incrementing or decrementing any register with a conditional branch depending on the result of the arithmetic operation that they carried out. These instructions increment or decrement the register f. If the resulting value is zero, a branch is produced. Otherwise, the program executes the next instruction. The result can be placed in W (f is not modified) or in the register f. The location of the result is specified by the parameter d in the instruction. These instructions also produce a short branch, similar to the instructions btfsc and btfss that have been previously described. These instructions do not alter the STATUS register.

The instructions incfsz f,d and decfsz f,d are very useful because when combined with the instruction goto they can be used to program loops or iterations in which the number of iterations is controlled by the value located in the register f.

Example 4.14

This example illustrates how to program a loop.

```
            movlw times
            movwf COUNTER
Loop:
            ;
            ; Place here instructions for loop.
            ;
            decfsz COUNTER, f    ; Does COUNTER = 0?, Yes - branch to End.
```

```
                   goto  Loop  ; No - do a new iteration.
        End:
                   ;
                   ;
```

In this example the register is called COUNTER. It initially stores the number of desired iterations that has been represented by the constant called "times." The instruction decfsz decrements and updates the value of COUNTER. When COUNTER is different than zero, the program executes the instruction goto Loop, thus starting a new iteration. When COUNTER reaches the value zero, the program branches to the address with the label End. This finishes the execution of this loop.

4.2.4 Bit Manipulation Instructions

Although there are only two instructions that specifically operate with bits, they are very powerful as they allow setting to 1 or 0 any bit in any register in the microcontroller. Table 4.5 shows these instructions and example 4.15 illustrates their use.

Example 4.15

Select bank in the data memory.
Bits RP1 and RP0 are used to select a specific data memory bank. Bank 1 is selected with RP1 = 0 and RP0 = 1. The following code can be used to set these bits:

```
bcf       STATUS, RP1
bsf       STATUS, RP0
```

The instructions btfsc and btfss, described earlier, can also be considered as instructions that manipulate bits, although they are also used for control transfer.

4.2.5 Other Instructions

Table 4.6 shows the rest of the instructions that do not fit into any of the previous classifications. The instruction nop does not execute any operation other than spending one instruction cycle in the microcontroller.

TABLE 4.5

Bit Manipulation Instructions

Mnemonic		Operation	Affects	Cycles
bcf	f, b	0 = > f<b>	—	1
bsf	f, b	1 = > f<b>	—	1

TABLE 4.6

Other Microcontroller Instructions

Mnemonic	Operation	Affects	Cycles
nop	No operation	—	1
clrwdt	0 = > WDT	TO#, PD#	1
sleep	go to low power consumption	TO#, PD#	1

The instruction clrwdt resets the watchdog timer, that is, it sets its timer to zero. Bits TO# and PD# in the STATUS register are both set to 1. TO# indicates watchdog overflow and PD# indicates that the microcontroller is in low-power consumption mode.

The instruction sleep activates the low-power consumption mode and resets the watchdog timer. The bit TO# is set to 1 and the bit PD# is set to 0.

4.3 Assembler Language Elements (for MPASM Assembler from Microchip)

Similar to other programming languages, assembler language has its own rules to write and combine words to create instructions. These rules make up the syntax for assembler language. This section describes the syntax for programming medium-end PIC microcontrollers.

4.3.1 Introduction

An assembler language program is a sequence of text lines. Each text line can be:

- An instruction from the microcontroller's instruction set
- A directive to the assembler
- A macroinstruction (macro)
- A comment
- A label
- A blank line

A *directive* is an instruction written in the source code, directed toward the assembler program instead of to the microcontroller. A program written in assembler language has a combination of instructions for the microcontroller and instructions to the assembler. These instructions give the assembler indications, such as definition of variables, the location of the program within the memory, and location of data memory.

A *macroinstruction*, or macro for short, is a user-defined instruction. The macro contains instructions from the microcontroller's instruction set and directives to the assembler. After the macro has been defined, when it appears in the code, the assembler will substitute the macro with the instructions used in its definition.

A *comment* is text whose objective is to help programmers by making it easier to read and understand the source code. Comments are preceded by a semicolon (;). When the assembler finds a semicolon in a line of code, it ignores everything in the rest of that line.

A *label* is the symbol that identifies a line in the source code representing the address of an instruction. Labels must be placed in column 1 and must be followed by a colon (:).

Example 4.16

The following code segment has two instructions: one for the microcontroller and one for the assembler.

```
data_5   equ      0xA8      ; Defines symbol data_5 and assigns a value.
prog1:
         movlw data_5       ; Instruction loads W register with the value in
                            ; the symbol data_5.
```

The symbol data_5 represents a constant whose value is assigned by the directive equ. The symbol prog1 is a label that represents the address in the program memory in which the instruction movlw data_5 is located. The first line of code is an instruction for the assembler language and the third line is an instruction for the microcontroller. Line 2 only contains a label while line 4 only contains a comment.

The lines that contain instructions for the microcontroller are structured in different fields. Some of these fields may be optional. Their syntax is as follows:

```
[label[:]]  mnemonic [operand 1] [, operand 2]  [;comment]
```

The fields inside brackets are optional. The different fields are separated with one or more blank spaces or with tabs. If the instruction has two operands, a comma separates them. Operands can be constants, symbols, or expressions.

Directives and the mnemonic can be written in lowercase or uppercase. For example, movlw and MOVLW are both correct and represent the same instruction.

TABLE 4.7

Constants and Syntax Used in Assembler Language

Constant	Syntax	Example	Value
Decimal	D*'decimal_number'* *.'decimal_number'*	D'167' .'167'	0x000000A7
Hexadecimal	H*'hexadecimal'* *0xhexadecimal* *hexadecimal*H	H'A7' 0xA7 0A7H	0x000000A7
Octal	O*'octal'* *octal*O	O'247' 247O	0x000000A7
Binary	B*'binary'*	B'10100111'	0x000000A7
ASCII	A*'ASCII_char'* *'ASCIII'*	A'Z' 'Z'	0x0000005A

Note: The letters D, H, O, and B are used to indicate the type of constant. They can be written in lowercase or uppercase.

Constants are values used in the program. Constants can be either numerical or ASCII. Numerical constants may be written in decimal, hexadecimal, octal, or binary. An ASCII constant is made up by the binary code that represents that ASCII character. The assembler treats constants as 32-bit binary numbers. When trying to write the constant into a shorter field, the constant is truncated. Table 4.7 shows some examples of syntax for constants in the PIC assembler language.

Example 4.17

All these instructions store the decimal value 167 in the W register:

```
movlw    .167
movlw    0a7h
movlw    247O
movlw    b'10100111'
```

Note how the hexadecimal constants must start with a digit in order to not be misunderstood as labels.

Numeric constants may be preceded by positive (+) or negative signs (–) to indicate positive or negative values. When there are no signs, the assembler assumes a positive value.

Assembler programs can also use symbols. A *symbol* is a set of up to 32 alphanumeric characters. Symbols must start with a letter or with the character (_). Symbols are used to name:

- Instruction addresses. In this case, the symbol is called a label
- Constant data

- Data memory registers, either special function registers or general purpose registers
- Bits in the data memory registers

All symbols except labels must be defined before the source code uses them, using among others the directives equ and set. From the microcontroller's point of view, the symbols used to name registers are data memory addresses. This is because the instructions make references to the register though their address in the data memory. The assembler codes the name of the variable using the address represented by that name. In example 4.18, if the variable REG1 appears in one instruction, the assembler will code it using the value 20h. The symbols that name register bits must be defined with values 0 to 7 as the PIC registers are 8 bits long.

Example 4.18

Consider the program segment shown below:

```
CONST1          equ   0xA5     ;Define symbol CONST1 and assign value
                               ;A5h.
REG1    equ   20h              ;Define symbol REG1 and assign value 20h.
BIT3    equ   3                ;Define symbol BIT3 and assign value 3.
        org   0x10             ;Memory address where program starts .
prog1:
        movlw CONST1           ;Set W register to A5h.
        movwf REG1             ;W stored in REG1.
        bcf   REG1, BIT3       ;Set to 0 bit 3 in REG1
```

In this program the symbol CONST1 represents an 8-bit constant with value A5h. REG1 is the general purpose register located in address 20h in the active memory bank. The value of the symbol REG1 is 20h. BIT3 is a constant with value 3. Here it is used as the name for bit 3 in the register REG1. The symbol prog1 is a label that represents the address where the program starts. Its value is 10h, as it is given by the directive org 0x10.

In the assembler language program used by the PIC microcontrollers, the names of the special function registers, as well as the names of their bits, are not predefined symbols. This means that the programmer must define the symbols used to name these registers and their bits. To ease the programming process, Microchip gives the definition file for each device. This is a text file containing the symbols used by the manufacturer to name the special function registers and bits in that specific device. For example, in the microcontroller PIC16F873, the file PIC16F783.INC contains the definitions and names for this device. Each PIC microcontroller has a similar file. The source code program in assembler must then include the specific file for the device in order to use the symbols defined in the file. Including the specific device file is done by using the directive #include as shown in Example 4.19.

Example 4.19

When working with the PIC16F84 microcontroller, the file P16F18.INC contains the definition of the names of the special function registers and their bits.

These names can be included in the source program by writing a line of code with the directive to the assembler to include the file as:

```
#include P16F18.INC.
```

It is necessary to write this line before using the names of the registers in the program. Once this directive is given to the assembler, it is possible then to refer to the special function registers and their bits using the names defined in the file.

The following is a partial listing of the P16F84.INC file. Microchip supplies similar files for each PIC microcontroller.

```
; P16F84.INC Standard Header File, Version 2.00 Microchip Technology, Inc.
; This header file defines configurations, registers, and other useful bits of
; information for the PIC16F84 microcontroller. These names are taken to match
; the data sheets as closely as possible.
; = = = = = = = = = = = = = = = = = = = = = = = = = = = =
; Register Definitions
; = = = = = = = = = = = = = = = = = = = = = = = = = = = =
W               EQU H'0000'
F               EQU H'0001'
;----- Register Files-----------------------------
INDF            EQU H'0000'
TMR0            EQU H'0001'
PCL             EQU H'0002'
STATUS          EQU H'0003'
FSR             EQU H'0004'
PORTA           EQU H'0005'
PORTB           EQU H'0006'
EEDATA          EQU H'0008'
EEADR           EQU H'0009'
PCLATH          EQU H'000A'
INTCON          EQU H'000B'
OPTION_REG      EQU H'0081'
TRISA           EQU H'0085'
TRISB           EQU H'0086'
EECON1          EQU H'0088'
EECON2          EQU H'0089'
;----- STATUS Bits --------------------------------
IRP             EQU H'0007'
RP1             EQU H'0006'
RP0             EQU H'0005'
NOT_TO          EQU H'0004'
NOT_PD          EQU H'0003'
Z               EQU H'0002'
DC              EQU H'0001'
C               EQU H'0000'
;----- INTCON Bits --------------------------------
GIE             EQU H'0007'
EEIE            EQU H'0006'
T0IE            EQU H'0005'
INTE            EQU H'0004'
```

```
RBIE            EQU H'0003'
TOIF            EQU H'0002'
INTF            EQU H'0001'
RBIF            EQU H'0000'
;----- OPTION Bits -------------------------------
NOT_RBPU        EQU H'0007'
INTEDG          EQU H'0006'
TOCS            EQU H'0005'
TOSE            EQU H'0004'
PSA             EQU H'0003'
PS2             EQU H'0002'
PS1             EQU H'0001'
PSO             EQU H'0000'
;----- EECON1 Bits -------------------------------
EEIF            EQU H'0004'
WRERR           EQU H'0003'
WREN            EQU H'0002'
WR              EQU H'0001'
RD              EQU H'0000'
```

4.3.2 Expressions, Operations, and Operators

Expressions are constants and symbols combined with arithmetic and logic operators. Expressions can be written using parentheses, similar to writing algebraic expressions. Expressions can be used in the field for operands and instructions. They are evaluated during the assembling process. The result of this evaluation becomes the value of the operand.

Example 4.20

Expressions are evaluated during assembling, not during the program execution. In the following segment of code, REG1 is a data memory register located in address 20h in one of the data memory banks.

```
REG1  equ     20h
      movwf   REG1 + 1
```

The expression REG1 + 1 does not increment the content of REG1, but the value of the symbol REG1. Therefore, the result is 21h. This result is obtained during the assembling process, so 21h becomes the operand of the instruction movwf. When the program is being executed, the instruction movwf will copy the content of the W register into register 21h in the active memory bank.

Operators are the symbols that indicate the mathematic and logic operations defined in assembler language.

4.3.2.1 Arithmetic Operators

Table 4.8 displays the arithmetic operators and the operations that they represent. Arithmetic operators include basic mathematic operations with symbols and constants. These operators can be used to build expressions

TABLE 4.8

Arithmetic Operators and Operations

Operator	Operation	Example
+	Addition	A1 + A2
–	Subtraction	A1 – A2
*	Multiplication	A1 * A2
/	Division	A1/A2
%	Mod (reminder in a division)	A1%A2

that become part of the operands in instructions or in directives to the assembler. Arithmetic operations use 32-bit registers, although their result can be truncated to the length of the register or address assigned to the expression.

Example 4.21

The following example illustrates the use of mathematical operators, in particular the operation mod.

```
DATA1    equ      .18
DATA2    equ      .7
;
         movlw DATA1%DATA2
```

The expression DATA1%DATA2 is evaluated during the assembling process using the values of the symbols DATA1 and DATA2 previously defined. The result of the operation mod is the residue of the division operation, in this case 4. This is the value that will be placed into the W register when the program is executed. Although the assembler uses 32 bits to calculate this expression, the result will be only the 8 least significant bits as this is the length of the W register.

The plus and minus signs can also be used to indicate if a constant is positive or negative. For example, the symbol "–" placed before a symbol or a constant generates the 2-complement for that symbol or constant using 32 bits When assigning this value to a shorter register, it will be truncated to the appropriate length using the least significant bits.

Example 4.22

The following segment of code defines the symbol DATA and assigns to it the value decimal 3. What is the value of the expression –DATA? What will be stored in W when the program is executed?

```
DATA equ     .3
     movlw -DATA
```

Expressions are evaluated using 32 bits Therefore, the value of the expression –DATA is FFFFFFFDh, which is the 2-complement representation of –3 using 32 bits When the instruction is coded, the assembler will use the 8 least significant bits, which is FDh. Therefore, during the execution of the program, the value in W will be FDh, which is the 2-complement representation of –3 using 8 bits.

4.3.2.2 Logic and Boolean Operators

Table 4.9 shows the logic and Boolean operators. The symbols and constants used in these operations can have only two values: TRUE or FALSE. A symbol or constant has the logic value TRUE if its numerical value is different than zero and FALSE if its numerical value is equal to zero.

TABLE 4.9

Logic and Boolean Operators

Operator	Operation	Example
!	NOT	! A1
&&	AND	A1 && A2
||	OR	A1 || A2
>	Higher than	A1 > A2
<	Less than	A1 < A2
>	Higher or equal to	A1 >= A2
<	Less or equal to	A1 <= A2
=	Equal to	A1 == A2
!=	Different than	A1 != A2

The evaluation of a logic or Boolean operation can only take the values TRUE or FALSE. If the logic value is TRUE that expression has the logic value 1; if the logic value is FALSE the expression has the logic value 0.

Example 4.23

Assume symbols A1 and A2 having the values 20h and 21h, respectively. What is the value of the different expressions? What is the value placed in the W register after executing the following instructions?

Expression	Result after Evaluation with A1 = 20h and A2 = 21h	Instruction Using the Expression	Value in W after Executing the Expression
! A1	FALSE	movlw ! A1	00h
A1 && A2	TRUE	movlw A1 && A2	01h
A1 || A2	TRUE	movlw A1 || A2	01h
A1 > A2	FALSE	movlw A1 > A2	00h
A1 < A2	TRUE	movlw A1 < A2	01h
A1 >= A2	FALSE	movlw A1 >= A2	00h
A1 <= A2	TRUE	movlw A1 <= A2	01h
A1 == A2	FALSE	movlw A1 == A2	00h
A1 ! = A2	TRUE	movlw A1 != A2	01h

4.3.2.3 Logic Operators Using Direct Bit Manipulation

Table 4.10 shows the logic operators that work directly with bits. These operators use the logic operations by directly modifying the bits of the symbols and constants. The operations AND, OR, and XOR take place among bits in the same position. The result of these operations is a binary number.

Example 4.24

Assume symbols A1 and A2 with values 3 and 5, respectively. What is the value of the different expressions? What is the value placed in the W register after executing the following instructions?

Expression	Result with A1 = 3 and A2 = 5	Instruction Using the Expression	Value in W after Executing the Expression
~ A1	FFFFFFFCh	movlw ~ A1 + 1	FDh (that is, -3)
A1 & A2	00000001h	movlw A1 & A2	01h
A1 | A2	00000007h	movlw A1 | A2	07h
A1 ^ A2	00000006h	movlw A1 ^ A2	06h
A1 >> 1	00000001h	movlw A1 >> 1	01h
A1 << 2	0000000Ch	movlw A1 << 2	0Ch

4.3.2.4 Assign Operators

Assign operators are used for assigning a value to a symbol (table 4.11). The assigned value can be the result of evaluating an arithmetic or logic

TABLE 4.10

Logic Operators Using Direct Bit Manipulation

Operator	Operation	Example
~	NOT	~ A1
&	AND	A1 & A2
|	OR	A1 | A2
^	Exclusive OR (XOR)	A1 ^ A2
>>	Right shift	A1 >> 1
<<	Left shift	A1 << 2

TABLE 4.11

Assign Operators

Operator	Operation	Example	Meaning
=	Logic or arithmetic assignment	var = 0	var = 0
++	Increment	var ++	var = var + 1
--	Decrement	var --	var = var − 1
+ =	Add and assign	var += k	var = var + k
- =	Subtract and assign	var -= k	var = var − k
* =	Multiply and assign	var *= k	var = var * k
/ =	Divide and assign	var /= k	var = var / k
% =	Mod and assign	var %= k	var = var % k
& =	AND and assign	var &= k	var = var & k
| =	OR and assign	var |= k	var = var | k
^ =	XOR and assign	var ^= k	var = var ^ k
>>=	Right shift and assign	var >>= k	var = var >> k
<<=	Left shift and assign	var <<= k	var = var << k

expression. These operators are used to define a symbol, simultaneously assigning an initial value to the symbol or modifying the value of a symbol previously defined. The directive set is to define symbols whose value can be changed during the assembling process.

Example 4.25

The following program code illustrates the use of assign operators. The program stores the values 10 and 15 in the registers 20h and 21h.

```
DATA = .10            ; Define symbol DATA. Store initial value 10.
REGISTER = 0x20       ; Define symbol REGISTER. Store initial value 20h.
        movlw DATA    ; Store 10 in W and
```

```
      movwf REGISTER    ; save it in register 20h.
DATA + = .5             ; Now, the value for symbol DATA is 15
REGISTER ++             ; and symbol REGISTER equals 21h.
      movlw DATA        ; Store 15 in W and
      movwf REGISTER    ; save it in register 21h.
```

The first two program lines define the symbols DATA and REGISTER with their initial values. The symbol DATA is used by the program as a constant and REGISTER represents the address of a data memory register. The values for both symbols are changed during program assembly. Therefore, instructions in the lines 7 and 8 are coded with operands different from the operands in lines 3 and 4.

4.3.2.5 Addressing Operators

Table 4.12 shows the addressing operators. These operators work with memory addresses. The operator $, when used as the instruction operand, signifies the real address for the instruction.

Example 4.26

The instruction

```
    goto  $
```

has the same effect as the instruction:

```
    prog: goto  prog
```

The operators low, high, and upper return bits 0 to 7, 8 to 15, and 16 to 21 from the label they operate. Because memory addresses in medium range PIC microcontrollers have 13 or fewer bits, the operator upper is not used in these microcontrollers. The operator high returns bits 8 to 12 from the address it operates. Example 4.10 (section 4.2.3.1) shows how to use the operators low and high.

TABLE 4.12

Addressing Operators

Operator	Operation	Example
$	Real address	goto $
low	Address low byte	movlw low label
high	Address high byte	movlw high label
upper	Address highest byte	movlw upper label

4.3.3 Directives

This section describes the most commonly used directives. The detailed description for all the MPASM directives is available on the Microchip Web site (http://www.microchip.com). *Directives* are instructions directed to the assembler program (instead of to the microcontroller) that will execute the program. A typical source code program written in assembler language has a mix of directives and instructions to the microcontroller. Directives are used to control the assembling operation, indicating different characteristics of the assembling process. For example, directives tell the assembler program the type of microcontroller that will be used; the definition of symbols used to name data, register, and bits; the initial address for the memory program; and so forth.

The general syntax for directives is as follows:

```
[label[:]] directive  [operands] [;comment]
```

If the directive has several operands, these are separated with commas. Operands can be constants, symbols, or expressions. The fields with brackets are optional. Table 4.13 shows the most commonly used directives. The full list of directives is available on the Microchip Web site.

TABLE 4.13

Most Commonly Used Directives By the Assembler MPASM for PICs

General Use Directives	
Goal of Operation	*Directives*
Define microcontroller and number system	list, processor, radix
Include a file within the source code, for example, a .INC file with definition of symbols	#include
Define symbols	equ, set,
Set program origin	org
Finish source code program	end

Directives Used for Relocatable Code	
Goal of Operation	*Directives*
Indicate beginning block of instructions	code
Indicate beginning block of data	udata, udata_shr
Reserve space in data memory	res
Indicate how symbols will be used	global, extern
Select page in program memory	pagesel
Select bank in data memory	banksel, bankisel

4.3.3.1 *General Use Directives*

Most programs written in assembler language use general directives. These directives give the assembler program information about:

- The type of microcontroller PIC for which the program is written
- The default number system
- The file that contains the definitions for the register and bit symbols
- The symbols used to name general purpose registers
- The address of the first line of program after which the assembler will have to code the program instructions.

4.3.3.1.1 *Directives list, processor, and radix*

The syntax for the directive list is:

```
list [option1] [,option2] [,…] .
```

The directive list turns listing output on and controls its format. Most of the options used for this directive do not control the format of the listing but the assembling. These options are shown in Table 4.14 and Example 4.27 illustrates the use of this directive.

Example 4.27

The MPASM assumes by default that numerical constants are in hexadecimal. It also assumes by default that the hexadecimal file that it generates will be in the Intel 8-bit standard format. If none of these parameters needs to be changed,

TABLE 4.14

Several Options for the List Directive

Option	Action
p = type of processor	Type of microcontroller. For example, p = 16f873 tells the assembler that the microcontroller used is a PIC16F873. This option does not assume a default value.
r = number system	Informs the numerical system that is used when writing a numeric constant in the program. (decimal, DEC; hexadecimal, HEX; octal, OCT). Example r = DEC. The default system is hexadecimal.
f = hex format	Specifies the format for the hexadecimal file:
	Standard 8-bit hexadecimal: INHX8m.
	8-bit separated hexadecimal: INHX8S.
	32-bit extended hexadecimal: INHX32.
	The default value is the standard 8-bit hexadecimal format.

the list directive will only have to specify the type of microcontroller. For a PIC16F873, the program line with the directive will be:

```
list  p = 16f873
```

It is necessary to keep in mind that the word "list" must be written after at least the second column in the line. Otherwise, it will be understood as a label, generating an error as the word "list" is a reserved word.

A different way to declare the type of microcontroller and the number system is by using the directives processor and radix. Their syntax are:

```
processor   type_processor
radix       number_system
```

Example 4.28 shows how to use these directives.

Example 4.28

We want to declare the microcontroller being used as a PIC16F84A, and the decimal numbering system will be used by default to write numerical constants. There are two ways to write the declarations. The first one uses the directives processor and radix:

```
processor  16f84a
radix      dec
```

The second way uses the directive list:

```
List       p = 16f84a, r = dec
```

Those constants written in the program without specifying their numerical system will be understood by the assembler as being written in the system declared by the directive radix or by the option r in the directive list.

Example 4.29

The following segment of code, based on an example shown in the help of the MPASM assembler, illustrates how the constants are interpreted depending on the radix and list directives.

```
list    r = dec ; From now on, constants are in decimal unless
                ; otherwise specified.
                ;
movlw   50H     ; This is 50 hexadecimal.
movlw   0x50    ; A different way of writing 50 hexadecimal.
movlw   50O     ; This is 50 octal.
```

```
movlw    50        ; This is 50 decimal because its number system
                   ; is not specified.
radix    oct       ; From now on, constants are in octal unless
                   ; otherwise specified.
                   ;
movlw    50H       ; This is 50 hexadecimal.
movlw    0x50      ; Another way of writing 50 hexadecimal.
movlw    .50       ; This is 50 decimal.
movlw    50        ; This is 50 octal because its number system
                   ; is not specified.
radix    hex       ; From now on, constants are in hexadecimal unless
                   ; otherwise specified.
                   ;
movlw    .50       ; This is 50 decimal.
movlw    50O       ; This is 50 octal.
movlw    50        ; This is 50 hexadecimal because its number system
                   ; is not specified.
```

4.3.3.1.2 Directives equ and set

The syntax for the directives equ (define constant) and set (define variable) are:

```
symbol equ  expression
symbol set  expression
```

These directives assign the value of the expression to symbol. They differ in the fact that the value of a symbol defined using the directive equ cannot be later modified by the assembler. However, the symbols defined using the directive set can be changed later in the program.

The directive equ is commonly used to define symbols associated with the microcontroller's hardware and for this reason will not change, such as the special function register names and their addresses in the data memory. It is also used to name constant data.

In programs designed to use absolute object code (therefore not using a linker), the directive equ is used to name the general purpose registers that are used in the program and to assign these names to their appropriate RAM addresses. For relative or relocatable object code, it is recommended not to use this directive to define general purpose registers. It is better to use the directive res within a data block declared with the directives udata or udata_shr.

Example 4.30

The following code segment illustrates the use of the directives equ and set.

```
REG1       equ  20h      ; REG1 is register 20h in data memory.
REG2       equ  21h      ; REG2 is register 21h in data memory.
DAT        set  .15      ; DAT is data with an initial value of 15.
           movlw DAT
           movwf REG1    ; Store 15 in REG1
DAT++                    ; Modify value of symbol DAT.
           movlw DAT
           movwf REG2    ; Store 16 in REG2.
```

4.3.3.1.3 Directive #include

The syntax for the directive #include (include additional source file) is:

```
#include file_name
#include "file_name"
#include <file_name>
```

in which file_name is the full name of a text file. If the name of the file contains blank characters, the first syntax mode cannot be used. This directive inserts the full text indicated in the specified file in the position in which the directive is located within the source code. The directive #include is usually used to insert the definition file containing the definition of the names for the special function registers and bits into the source code. With this, the user does not have to declare these names in the source code.

Example 4.31

It is common to declare the type of processor and the names of the special function register and bits at the beginning of the source code written in assembler. This is done by using the directives list and #include. The following piece of code illustrates how to write these directives if the selected microcontroller is PIC16F873 and definition file is named P16F873.INC:

```
List        p = p16f873
#include    p16f873.inc
```

4.3.3.1.4 Directive org

The syntax for the org (origin of program) is:

```
[label]   org   expression
```

This directive sets the program origin for subsequent code at the address defined by the expression. If the directive uses a label, the label receives the value of the expression.

Example 4.32

The directive org is normally used to indicate to the assembler program the memory address that corresponds to the reset (address 0) and to the interrupt vector (address 4). The following code shows an example of this:

```
list        p = p16f873
#include    p16f873.inc
org         0           ; Set address 0.
movlw       high PP     ; This instruction is set at address 0.
```

```
        movwf       PCLATH      ; This instruction is set at address 1.
        goto        PP          ; This instruction is set at address 2.
        org         4           ; Set address 4;
        ;
        ; Write here instructions for interrupt subroutines
        ; that will be assembled after address 4.
        ;
        ;
        ; Main program will be written starting at address 800h.
        ;
PP:     org   800h              ; Set address 800h.
        ;
        ; Write here the main program.
        ;
```

4.3.3.1.5 Directive end

The syntax for the directive end (finish source code) is: end. This directive tells the assembler program to finish assembling the source program. This directive is placed in the last line in the source program. The assembler will ignore anything after directive end.

4.3.3.2 Directives for Relocatable Code

This section describes the most used directives when writing relocatable code. These are programs that need a linker in addition to the assembler program because the source code does not specify absolute addresses. These directives are used to:

- Symbolically indicate the beginning of data or instruction blocks
- Reserve space in the data memory for the variables used in the program
- Indicate which are global or external symbols in programs with several source code files
- Easily select pages in the program memory or register banks in the data memory

4.3.3.2.1 Directive code

This directive indicates the beginning of a section of program code. Its syntax is:

```
[label]   code [ROM_address]
```

The field ROM_address indicates the address for the beginning of the section or group of instructions used in relocatable code. If the field ROM_address does not exist, the linker decides the initial address. The field label is used to name the section.

Two sections cannot have the same name. If the field label is not used, the section receives the name .code. A section of code ends when another section of code begins, or when it reaches the directive end.

Example 4.33

The following code segment illustrates how to use the directive code to declare sections of code.

```
Rst_vector      code  0       ; This section begins at address 0.
   movlw        high PP
   movwf        PCLATH
   goto         PP
Intr_vector     code  4       ; This section beings at address 4.
   goto         SR_Int
Intr_Prog       code  5       ; This section begins at address 5.
SR_Int:
;
; Write here instructions for subroutine interrupts.
;
Prog_Principal  code          ; The linker will set the initial
                              ; address for this section.
PP:
;
; Write here the main program.
;
```

4.3.3.2.2 Directives udata, udata_shr, and res

The directives udata (being a section of uninitialized data) and udata_shr (being a shared section of initialized data) use the following syntax:

```
[label]     udata       [RAM_address]
[label]     udata_shr   [RAM_address]
```

These directives are used to declare the beginning of data sections. The label RAM_address specifies the first data memory address. If this label does not exist, the linker decides the initial address.

The directive udata is used for register sections located in a single data memory register bank. The directive udata_shr is used to declare sections that share more than a single memory bank. For example, in the PIC16F873 the registers located in address 20h to 7Fh in bank 0 are repeated at the same addresses in bank 2 (Figure 3.11). Therefore, when referring to this group of registers it is necessary to use udata_shr instead of udata. The term "uninitialized data" means that there is not an initial value to data in this section when they are defined. The directive res is used to define uninitialized data. The syntax for the directive res (reserve data memory) is:

```
[label]   res memory_size
```

This directive forces the memory counter to advance in memory_size number. It is used to separate space in the data memory without assigning an initial value to this space. It is used within the data sections declared with the directives udata and udata_shr.

Example 4.34

The following section of code illustrates the use of the directives udata_shr and res.

```
        udata_shr
REG1    res         1
REG2    res         1
```

This declares a shared section of uninitialized data with the symbols REG1 and REG2, reserving a memory cell for each one of them. The linker will assign the addresses for these registers.

4.3.3.2.3 Directives global and extern

The syntax for the directives global (export a symbol) and extern (declare an externally defined symbol) are as follows:

```
global   symbol [, symbol...]
extern   symbol [, symbol...]
```

These directives are used when there are several modules that need to be linked, with symbols defined in one module and used in another one. The directive global declares symbols defined in one module that must be available in other modules. The directive extern declares symbols that are used in the current module but have been defined in another module, in which they have been defined using the directive global.

Example 4.35

The following project consists of two modules. The first module contains the main program in the file named pp.asm. The second module, called sr.asm, contains the subroutines called from the main program. The subroutine Delay, one of the subroutines defined in this module, produces a delay that is proportional to the value stored in the register named with the symbol REG. The main program calls this subroutine and gives it the delay value located in the register REG.

For the assembling and linking process to be correct, the main program declares the symbol Delay externally and defines the symbol REG as part of the data section, declaring it as a global symbol. On the other hand, the subroutine module defines the symbol Delay as global inside a program section and the symbol REG as external.

Main program module (file named pp.asm):

```
;
; Main program module.
; Examples of using directives global and extern.
;
        list       p = 16f873
        #include   p16f873.inc
        udata_shr
REG     res        1          ; Define symbol REG.

        global     REG        ; Symbol REG declared as global in this
                              ; module
                              ; and external in the subroutine module.
        extern     Delay      ; Symbol Delay defined external in this module
                              ; defined and declared global in subroutine
                              ; module.
;
; The call to subroutine Delay is in one of the sections in the main
  program (for example, in the Program Section).
;
;
Program            code
;
        movlw      35h
        movwf      REG
        call       Delay
;
        end
```

Subroutine module (file sr.asm):

```
;
; Subroutine program module.
; Examples of using directives global and extern.
;
        list       p = 16f873
        #include   p16f873.inc
        global     Delay      ; Symbol Delay defined global in this module
                              ; and external in the main program module.
    extern         REG        ; Symbol REG declared external in this module
                              ; and global in the main program.
;
; The call to subroutine Delay is in one of the sections in the main
  program.
;
Program code
Delay:
        decfsz     REG,1
        goto       Delay
        return
;
        end
```

4.3.3.2.4 *Directives pagesel, banksel, and bankisel*

The syntax for the directives pagesel (select memory page), banksel (direct selection of register bank), and bankisel (indirect selection of register bank) are as follows:

```
pagesel label
banksel label
bankisel label
```

The directive pagesel produces the necessary code to select the memory page where the label specified in the directive is located. This directive introduces the necessary instructions in the program to modify the register PCLATH. If the microcontroller has only one memory page, this directive does not generate any code.

The directives banksel and bankisel produce the necessary code to select the memory bank in which the label specified in the directive is located, using direct or indirect addressing. The directive banksel is equivalent to introducing instructions to manipulate bits RP1 and RP0 in the STATUS register. The directive bankisel manipulates the bit IRP in the STATUS register to assign it the appropriate value.

Example 4.36

The following program illustrates the use of the directives pagesel, banksel, and bankisel.

```
        list        p = 16f873
        #include p16f873.inc
; Constant data:
DATA1 equ          0x55
DATA2 equ          .10
; Data memory registers:
  udata_shr
REG1    res         1              ; REG1 is register 20h in the data memory.
REG2    res         1              ; REG2 is register 21h in the data
; memory.
; Programs:
Rst_vector          code  0
        pagesel     PP             ; Select page where PP is located.
        goto        PP             ; This guarantees branching to
                                   ; correct address.
Prog_Principal      code
PP:
        pagesel     SRoutine       ; Select page where subroutines are
                                   ; located
        call        SRoutine       ; guaranteeing branching to correct
                                   ; location.
;
; Operate with registers TRISB and PORTB using direct address:
        banksel     TRISB          ; Select bank 1 because TRISB is in
                                   ; this bank.
                                   ; This assures the correct
                                   ; addressing for TRISB.
```

```
        clrf       TRISB        ; Work with TRISB.
        banksel PORTB           ; Return to Bank 0, because PORTB is
                                ; located in this bank.
        movf       PORTB, DATA1 ; Work with PORTB.
;
; Operate with REG1 using indirect address:
        movlw      REG1
        movwf      FSR          ; Stored address of REG1 in FSR.
        bankisel   REG1         ; Select bank where REG1 is located.
        movlw      DATA2        ; Write 10 in
        movwf      INDF         ; REG1 using indirect addressing.
SRoutine:
;
; Write here subroutine instructions.
;
        return
        end
```

4.3.4 Macroinstructions

Macroinstructions (or macros for short) are instructions defined by the user using microcontroller instructions and assembler directives. Once a macro has been defined, it is possible to call it from the source code. Macroinstructions are defined using the following syntax:

```
macro_name macro   [arg_def1, arg_def2,...]
           [ local  label [, label, label,...]]
;
; Body of macroinstruction
;
              endm
```

In this syntax, macro_name is the symbol for the name given to the macroinstruction, and arg_def1, arg_def2, and so forth are optional arguments used in the definition. These arguments are symbols used in the body of the macro. The first line in the definition contains the directive macro, declaring the name of the macro (macro_name) and the arguments (arg_def1, arg_def2, etc.) if these exist. The last line in the macro is used by the directive endm, which tells the assembler that this is the end of the macro. Each directive macro must have a directive endm.

The body of the macro contains the instructions and directives that program the algorithm the programmer has decided to group as a macroinstruction. The body of the macro uses the arguments declared in the directive macro as part of the operands for the instructions and directives. It is common for the body of the macro to contain local labels. These labels must be declared in the body of the macro using the directive local. Once a label has been declared as local, it does not matter if there is another label with the same name outside the macro.

The macro is called by writing its name and arguments within a line of program:

```
Macro_name [arg1, arg2,...]
```

The assembler places the body of the macro in the source code, where it was called. When expanding the directives and instructions, the assembler uses the arguments arg1, arg2, ... to substitute the symbols used in the definition of the macro. Arguments can be either symbols or expressions.

Example 4.37

The following program defines the macro Convert. This macro is called twice by the program. The macro receives a hexadecimal number in a register called HEXA and outputs the equivalent ASCII character in a register called ASCII. For example, if HEXA = 0Ah, then ASCII = 41h. The conversion to ASCII is done by adding 30h to the hex number if this is less than or equal to 9, or by adding 37h if the hex number is higher than 9.

```
    list        p = 16f873
    #include    p16f873.inc
;
; Macro definition:
;
; This macro converts an hexadecimal digit (0 to F) located in
; register called HEXA
; into its equivalent ASCII character. The ASCII digit is then
; stored in the register called
; ASCII.
;
Convert macro HEXA, ASCII             ; Declare macro
        local       add30, add37, end_mac ; local labels.
        movf        HEXA, W           ; Store hex digit in HEXA.
        sublw       9                 ; W > 9 ? (affects
                                      ; STATUS<C>).
        movf        HEXA, W           ; Store hex digit in W
                                      ; (does not affect STATUS<C>).
        btfsc       STATUS, C         ; Yes (C = 0), add 37h to HEXA.
        goto        add30             ; No (C = 1), add 30h to HEXA.
add37:
        addlw       37h
        goto        end_mac
add30:
        addlw       30h
end_mac:
        movwf       ASCII             ; Store result in register ASCII.
        endm                          ; End of macro.
;
; Data memory registers:
;
        udata_shr
HEXA1 res       1
HEXA2 res       1
ASCII1   res    1
ASCII2   res    1
```

```
;
; Programs
;
Rst_vector      code  0
      pagesel   MP                      ; Select page where MP is
; located
      goto      MP                      ; to guarantee correct
; branching.
Main_Program:code   0x800
MP:
      movlw   9
      movwf   HEXA1
      movlw   0Ah
      movwf   HEXA2
;
      Convert HEXA1, ASCII1             ; Call macro Convert.
                                        ; Assembler program will
                                        ; introduce here
                                        ; the instructions for the
                                        ; macro.
;
      nop
;
      Convert HEXA2, ASCII2             ; Call macro Convert again.
                                        ; Assembler will introduce
                                        ; here the
                                        ; instructions for the
                                        ; macro again.
      nop
      goto  $                           ; Infinite loop.
      end                               ; End of program.
```

4.3.5 Organization of a Program in Assembler Language

Although there are no strict rules for writing a program in assembler language, it is recommended to write it in the following order:

1. Define the processor and its symbols by using the directives list and #include.
2. Write the definition of the macroinstructions that will be used in the program.
3. Define the symbols that will represent constant data with the directives equ and set.
4. Define the use of the data memory. This means to define the symbols used in the program to represent general purpose registers and their addresses.
5. Write the body of the main program. In general it should start by initializing the variables that require initial values.
6. Write subroutines.
7. Finish the program using the directive end.

It is important to note that the definition of symbols and their values, either data or addresses, is done before writing program instructions. The method for defining the location of the instructions in the program memory and the data in data memory is dependent on the use of absolute vs. relocatable code. In absolute coding, the assembler must have all the information necessary to code the source program. This requires all addresses used in the source program to be defined from the beginning. This includes the addresses where the blocks of instructions start, as well as the addresses for the general purpose registers used by the program. Program addresses are defined using the directive org. Register addresses are defined using the directive equ.

When using relocatable code, the assembler creates an incomplete coding of the source program because the absolute addresses are not specified. The linker is the program that sets the final addresses for memory and data. The body of the program can be organized in sections, each one of them starting with the directive code. Some of these sections, for example for the reset and interrupt vectors, must be placed in fixed sections in the program memory. In this case, it will specify the beginning address. For other sections that can be placed anywhere in the program memory, the linker will select the memory locations. Something similar happens with the data memory. Here, only the names of the registers used in the source program are declared, reserving memory size without specifying its address. This is done with the directives udata, udata_shr, and res.

Examples 4.38 and 4.39 show the recommended structure for source programs that will be assembled using absolute and relocatable code, respectively. The reader should analyze these two programs carefully.

Example 4.38

The following program shows the organization of the source code in assembler when using absolute code.

```
list        p = 16f873      ; Declaring the microcontroller to be
                            ; used
#include    <p16f873.inc>   ; and its variables
;
; Define constants:
;
DATA1           EQU   0x1     ;
DATA2           EQU   0x2     ;
;
; Define variables:
;
w_temp          equ   0x20    ; Variable used to store W.
status_temp     equ   0x21    ; Variable used to store STATUS.
X               equ   0x22    ; Example.
Y               equ   0x23    ; Example.
;
; Body of program:
```

```
;
    org     0x000           ; Reset vector address.
    movlw   high PP         ; Prepare branch to main program,
    movwf   PCLATH          ; guaranteeing correct address.
    goto    PP              ; Go to address where main program
                            ; starts.
;
    org     0x004           ; Interrupt vector address.
    movwf   w_temp          ; Save current content of W.
    movf    STATUS, W       ; Copy current STATUS in W,
    bcf     STATUS, RP0     ; Assure selection of bank 0
    movwf   status_temp     ; and save content in STATUS
;
; Write here subroutines for interrupt requests.
;
    bcf     STATUS, RP0     ; Assure selection of bank 0.
    movf    status_temp, W  ; Recall copy of STATUS
    movwf   STATUS          ; and write it back.
    swapf   w_temp, f       ; Recall copy of W
    swapf   w_temp, W       ; and write it back without altering
                            ; STATUS.
    retfie                  ; Return from interrupt.
;
PP:
    clrf    X               ; Initialize variables.
    clrf    Y               ; Initialize variables.
;
; Write here instructions of the main program.
;
    movlw   high SR1        ; If SR1 is in a different page
    movwf   PCLATH          ; guarantee selection of correct page
    call    SR1             ; and call subroutine.
;
; Write here instructions of the main program
;
    goto    $               ; Example: infinite loop to finish
                            ; main program.
;
SR1:                        ; Beginning of subroutine SR1.
;
; Write here instructions for subroutine SR1.
;
    movlw   high SR2        ; Example: Call subroutine 2 that is
                            ; in another page
    movwf   PCLATH          ; guarantee selection of correct page
    call    SR2             ; and call subroutine SR2.
;
; Write here instructions for subroutine SR1.
;
    return                  ; Return to main program from SR1.
SR2:                        ; Beginning of subroutine R2.
;
; Write here instructions for subroutine SR2.
;
    return                  ; Return to SR1 from SR2.
    end                     ; End of source program.
```

Example 4.39

The following program shows the organization of the source code in assembler when using relocatable code.

```
        list      p = 16f873      ; Declaring the microcontroller to be
        #include <p16f873.inc>    ; used and its variables.
;
; Define constants:
;
DATA1         equ    0x1          ; Example.
DATA2         equ    0x2          ; Example.
;
; Define variables:
;
    udata_shr
w_temp        res    1            ; Variable used to store W.
status_temp res    1            ; Variable used to store STATUS.
X             res    1            ; Example.
Y             res    1            ; Example.
;
; Body of program:
;
Rst_vector code  0               ; Reset vector in address 0
    pagesel PP                    ; Prepare branch to main program and
    goto    PP                    ; branch to address where it starts.
;
Intr_vector code 4               ; Interrupt vector in address 4
    goto    SR_Int                ; Branch to interrupt subroutine.
;
Intr_Prog  code 5               ; Section with interrupt subroutine.
SR_Int:
    movwf    w_temp               ; Save current content of W.
    movf     STATUS, W            ; Copy current content of STATUS in W.
    bcf      STATUS, RP0          ; Assure selection of bank 0 and
    movwf    status_temp          ; save content of STATUS.
;
; Write here instructions for interrupt subroutine.
;
    bcf      STATUS, RP0          ; Assure selection of bank 0.
    movf     status_temp, W       ; Recall copy of STATUS
    movwf    STATUS               ; and write it back.
    swapf    w_temp, f            ; Recall copy of W.
    swapf    w_temp, W            ; Write it back without altering
                                  ; STATUS.
    retfie                        ; Return from interrupt.
;
Prog_Principal    code           ; Write main program here.
PP:
    clrf     X                    ; Initialize variables.
    clrf     Y                    ; Initialize variables.
;
; Write here instructions for main program.
;
    pagesel SR1                   ; Select page where SR1 is located.
    call    SR1                   ; Call subroutine SR1.
;
; Write here instructions for main program1.
;
    goto $                        ; Example: infinite loop.
```

```
;
Subroutines code              ; Section for subroutines.
SR1:                          ; Beginning of subroutine SR1.
;
; Write here instructions for subroutine SR1.
;
    pagesel SR2               ; Example: From subroutine SR1
    call    SR2               ; call subroutine SR2.
;
; Write here instructions for subroutine SR1.
;
    return                    ; Return to main program.
;
SR2:                          ; beginning of subroutine SR2.
;
; Write here instructions for subroutine SR2.
;
    return                    ; Return to subroutine SR1.
;
    end                       ; End of source code.
```

Macroinstructions and subroutines are two very valuable programming resources that permit the use of a modular approach when writing the source code program. They allow the programmer to write an algorithm only one time, but use it as needed, each time with different parameters. In addition to reducing programming time, they are also helpful in detecting errors and making programs easier to understand. Subroutines and macroinstructions differ on when they are invoked. Macroinstructions are invoked or called during the assembling stage, whereas subroutines are called during the execution of the program.

When the assembler finds a call for a macroinstruction, it inserts in that space the instructions that define the macro. Therefore, the size of the program memory being used will increase as the number of calls to the macroinstruction increases.

Subroutines are defined only once and they are placed by the assembler or linker somewhere in the program memory. Subroutines are called during the execution of the program. The effect of the subroutine call is to branch the program toward the instructions that define the subroutine, executing them in the same order they were written. The program returns to the point after it was branched by the call to subroutine. The number of times a subroutine is called does not have any effect on the size of the program.

The execution of a subroutine requires three elements: instruction for calling the subroutine, instruction for returning, and the stack. The main objective of the stack is to store the address for returning to the program from which the subroutine was called. The instructions for calling the subroutine and return, because they have to use the stack, increase the execution time. However, this increase in the execution time is normally minor.

An additional advantage of working with subroutines is the possibility to build and use subroutine libraries. A library is a file that contains a collection of programs (subroutines) generally on a specific theme. During the linking process, the linker takes from the library only those subroutines requested by the program and incorporates them within the object program.

Choosing between using macroinstructions or subroutines depends largely on the programmer's preferences. However, for those microcontrollers with limited program memory, it can be better to use subroutines instead of macroinstructions if the algorithm is called several times.

4.4 Available Resources for Programming PIC Microcontrollers in Assembler Language

The programs used to program applications in assembler language in medium-end PIC microcontrollers are:

Text Editor: This program is used to create the source program (file. asm).

Assembler (MPASM.EXE or MPASMWIN.EXE): This program translates the source code file (in the file.asm) into machine language. The translation may be full or partial. If the translation is full, the assembler produces a hexadecimal file (.hex) that contains the coded program. If the translation is partial, it produces an object file (.o) that will become one of the inputs to the linker program.

Linker (MPLINK.EXE): This program creates the machine language file (.hex) by linking the different object modules (files .o) produced by the assembler and library files in a single module.

Library manager (MPLIB.EXE): This program creates a library (.lib) from several assembler programs. A library is a collection of programs.

Simulator/Debugger: This program simulates the microcontroller in a personal computer. It has several orders that are used to test the programs and debug errors. This program is included in the integrated development environment MPLAB.

Programmer: This program is used, with some additional hardware, to program the microcontroller. The programmer takes the hexadecimal file (.hex) produced by the assembler or the linker, then programs it into the device.

Integrated Development Environment MPLAB: This is an integrated toolset for the development of applications using PIC microcontrollers, including the editor, assembler, linker, and debugger. With the use of the appropriate hardware it can also be used to program the microcontroller.

4.4.1 The MPASM Assembler

The MPASM translates the source program in the file.asm into machine language. When using a personal computer, the MPASM assembler can be used as a stand-alone program from the DOS prompt line (MPASM.EXE), from the Windows operating system (MPASMWIN), or from the Integrated Development Environment. Figure 4.9 shows the assembling process and its associated files.

The assembler can generate absolute or relocatable code depending on whether a full or complete translation of the program source is performed. When doing a full translation, the assembler will produce a hexadecimal file (.hex). When doing a partial translation the assembler will produce an object file (.o) that will be used as the input to the linker MPLINK.

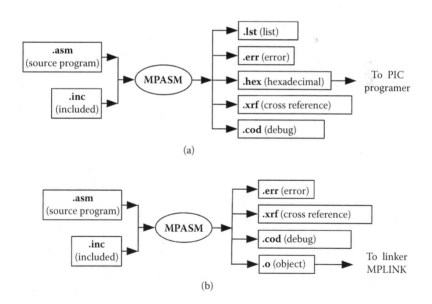

FIGURE 4.9
Assembling process with MPASM and files involved in this process. (a) Assembling with absolute codification. The hexadecimal file (.hex) contains the program translated into machine language and ready to be stored in the microcontroller's memory. (b) Relocatable code assembling. It has partial translation into machine language. The output file is an object file (.o) that must be processed with the linker MPLINK.

4.4.1.1 Absolute Code Generation

In absolute code generation, the program MPASM carries out the process for translating the source code file into machine language. The result of this process is a file with the program fully codified. Absolute code generation can only be done when the source program contains all the information that the assembler needs to translate it into machine language. This means that the source program contains the specifications for the memory addresses to store instructions and the addresses in memory for the registers used by the program. With this information, the assembler is able to fully codify the source program.

Absolute code generation requires the source program be totally contained within a source file (.asm) or with parts of the program contained in files that will be included during the assembly process with the directive #include. As a result of the assembling process, the MPASM program generates the machine language code in hexadecimal format (.hex). The programmer can then use this file to program the microcontroller.

4.4.1.2 Relocatable Code Generation

In relocatable code generation, the MPASM program creates a partial codification of the source program. This task will later by completed by the linker MPLINK. When the addressing information in the source program is not complete, the assembler can only carry out a partial and incomplete codification of the source code. The linker MPLINK will finish the process by generating the machine language program.

The result of the assembling process is then an object file (.o) that will be used as the input file to the linker MPLINK. The linker can receive several object files that will be linked to generate a single machine language program in hexadecimal format. The programmer will use this file to program the microcontroller.

4.4.1.3 Files Used and Generated during the Assembling Process

There are several files used during the assembling process. The assembler needs some of these files as input files, while other files are generated as output files. The files generated during the assembling process are fully described in the help areas for the Integrated Development Environment (IDE) MPLAB available on the Microchip Web site. The most important files involved in the assembling process are:

Source file (.asm): This is a text file that contains the source code. It can be written on a personal computer using any text editor, such as the editor included in the IDE MPLAB. This file is an input file to the assembler MPASM.

Include source file (.inc): This is a text file that contains part of the source program. This file is included in the source file using the directive #include. This file is mainly used to define the names and addresses for the registers and bits of the microcontroller that will be used. Microchip supplies this file for each one of the PIC microcontrollers that they manufacture. For example, the file 16f873.inc contains all the definitions for the names of the special function registers, their addresses, and the names of the bits in the PIC16F873 microcontroller. To use this file, the programmer has to include this file in the source code, as shown in example 4.19.

List file (.lst): This is a text file that contains the list of the source program, addresses, and the codification of its instructions. This file can be generated by the assembler MPASM or the linker MPLINK.

Object file (.o): This is a file produced by the assembler as a result of relocatable codification. It contains the partial codification of the source program. This file becomes one of the input files to the linker MPLINK. These are not text files.

Hexadecimal file (.hex): This is a text file that contains the instruction codes and their addresses in the Intel hexadecimal format. This file can be generated by MPASM (absolute codification) or MPLINK (relocatable codification). This file is the final product of the assembler process or the assembler and linking process.

This hexadecimal file contains a set of text lines called records using the following format: **:LLAAAATTDDDD...DDSS**.
The structure of a record is as follows:

:—ASCII character that indicates the beginning of a record.

LL—Length of the record. These are ASCII characters that indicate the length in hexadecimal values of the data contained in the record.

AAAA—Record initial address. These are four ASCII characters that represent the address of the first data (byte). Because medium-end PICs have program memory cells of 14 bits and their content is in 2 bytes, the address in this field is twice the real address.

TT—Indicates type of record. 00 indicates a data record. 01 indicates the last record in the hexadecimal file.

DD—This field contains the data. Two ASCII characters in this field represent the hexadecimal value of data (byte). Because medium-end PICs have program memory cells of 14 bits, the content of each cell needs four ASCII characters in this field.

SS—Checksum. This is calculated by adding all the bytes (not the ASCII codes) of the record. SS is then the 2-complement of this sum using 8 bits.

Example 4.40

This example shows the structure of the list and hexadecimal files obtained after assembling a very basic program. The list file is only partially shown.

Source file (example.asm):

```
    list        p = 16f873
    #include    <p16f873.inc>
X           equ         0x20
Y           equ         0x21
    org         0x000       ; Reset vector address.
    movlw       high MP
    movwf       PCLATH
    goto        MP .
    org         0x004       ; Interrupt vector address.
    retfie
    org         0x0123
MP:
    clrf        X           ; Sets X = 0.
    clrw
    addlw       1
    movwf       Y           ; Sets X = 1.
    end                     ; End of source code.
```

List file (example.lst) (partially shown)

```
LOC OBJECT CODE LINE SOURCE TEXT
                    VALUE
     00001              list      p = 16f873
     00002              #include <p16f873.inc>
     00001          LIST
     00002          P16F873.INC Standard Header File, Version 1.00
                    Microchip Technology, Inc.
     00358          LIST
     00003
  00000020  00004   X          equ      0x20
  00000021  00005   Y          equ      0x21
     00006
  0000      00007              org      0x000 ; Reset vector address.
0000 3001   00008              movlw    high MP
0001 008A   00009              movwf    PCLATH
0002 2923   00010              goto     MP
     00011
0004        00012              org      0x004 ; Interrupt vector
                                                   address.
0004 0009   00013              retfie
     00014
0123        00015              org      0x0123
0123        00016   MP:
0123 01A0   00017              clrf  X      ; Sets X = 0.
0124 0103   00018              clrw
0125 3E01   00019              addlw 1
0126 00A1   00020              movwf Y      ; Sets X = 1.
     00021
     00022              end              ; End of source code.
```

Hexadecimal file (example.hex)

```
:0600000001308A002329F3
:020008000900ED
:08024600A0010301013EA1002B
:00000001FF
```

4.4.2 The Linker MPLINK

The linker MPLINK writes the program in machine language using a hexadecimal format (.hex). The linker uses the following files:

- One or more object files (.o) produced by the assembler
- One or more library files (.lib) produced by the library manager MPLIB
- An auxiliary file (.lkr) that contains the description of the available memory for the microcontroller or any other data needed for the linking process

Figure 4.10 illustrates the linking process and the files involved in the process. When using a personal computer, the linking process can be executed as a stand-alone process (MPLINK using the command line in DOS) or as part of the IDE MPLAB.

The auxiliary file (.lkr) informs the linker of the addresses for the program and data memory in the selected microcontroller. This allows the linker to assign the correct addresses to finish the whole process. It can also contain information needed for the linking, such as the names for the files to be linked if those were not specified previously. The full

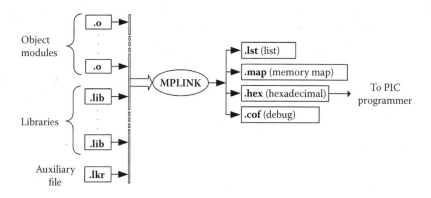

FIGURE 4.10

Linking process with MPLINK and files involved in this process. The object files (.o) are the output files from the assembler as a result of relocatable codification of the modules in assembler language. The library files (.lib) are obtained with the library manager MPLIB. The auxiliary file (.lkr) contains the description of available memory in the microcontroller so the linker can use the correct addresses. The hexadecimal file (.hex) contains the program translated into machine language and ready to be stored in the microcontroller.

description of the directives to MPLINK is available in the help section of the IDE MPLAB on the Microchip Web site.

The available memory addresses are described in the auxiliary file using the linker directives databank, sharebank, and codepage. The directive databank specifies the name, and the initial and final addresses of the region in the data memory that contains a bank of registers. The directive sharebank does the same but with respect to a region that shares addresses in more than one register bank.

The directive codepage specifies the name, and initial and final addresses for a program memory region. As an option it is possible to order the linker to totally fill this memory region with data (14 bits in medium-end PICs). The region can also be declared as protected, thus giving some restrictions in its use from the source code.

The syntax for these directives is:

```
databank  name = name start = init_addr end = end_addr [protected]
sharebank name = name start = init_addr end = end_addr [protected]
codepage  name = name start = init_addr end = end_addr [protected]
  [fill = value]
```

The connection between the memory sections declared using these variables and the source program sections is done using the directive section. This directive allows for declaring a section of memory as a logic section. Its syntax is:

```
section   name = section_name ROM = name
section   name = section_name RAM = name
```

In these directives, section_name is the symbol that identifies the section, and name is the name of the memory region that can be program memory (ROM) or data memory (RAM), previously declared using the directives databank, sharebank, and codepage. Logic sections can be used in the source code program by means of the directives udata, udata_shr, and code.

Example 4.41

This example shows the content of the auxiliary file that contains the description of the memory regions in a PIC microcontroller. It also shows how to use in the source code the logic sections declared with the auxiliary file 16f873.lkr that is provided by Microchip as part of the IDE MPLAB.

```
// Sample linker command file for 16F873
// $Id: 16f873.lkr, v 1.5 2002/11/07 23:16:07 sealep Exp $
LIBPATH.
CODEPAGE   NAME = vectors   START = 0x0   END = 0x4   PROTECTED
```

```
CODEPAGE  NAME = page0  START = 0x5     END = 0x7FF
CODEPAGE  NAME = page1  START = 0x800   END = 0xFFF
CODEPAGE  NAME = idlocs START = 0x2000  END = 0x2003PROTECTED
CODEPAGE  NAME = config START = 0x2007  END = 0x2007PROTECTED
CODEPAGE  NAME = eedata START = 0x2100  END = 0x217FPROTECTED
DATABANK  NAME = sfr0START = 0x0        END = 0x1F  PROTECTED
DATABANK  NAME = sfr1START = 0x80       END = 0x9F  PROTECTED
DATABANK  NAME = sfr2START = 0x100      END = 0x10FPROTECTED
DATABANK  NAME = sfr3START = 0x180      END = 0x18FPROTECTED
SHAREBANK NAME = gpr0START = 0x20       END = 0x7F
SHAREBANK NAME = gpr0START = 0x120      END = 0x17F
SHAREBANK NAME = gpr1START = 0xA0       END = 0xFF
SHAREBANK NAME = gpr1START = 0x1A0      END = 0x1FF
SECTION NAME = STARTUP   ROM = vectors  // Reset and interrupt
                                           vectors
SECTION NAME = PROG1     ROM = page0    // ROM code space
                                           - page0
SECTION NAME = PROG2     ROM = page1    // ROM code space
                                           - page1
SECTION NAME = IDLOCS    ROM = .idlocs  // ID locations
SECTION NAME = CONFIG    ROM = .config  // Configuration bits
                                           location
SECTION NAME = DEEPROM   ROM = eedata   // Data EEPROM
```

There are several sections declared in this auxiliary file. For example, the section named PROG1 corresponds to the program memory region named page0. This section starts in address 005h and ends in address 7FFh.

It is possible in the source program to indicate the linker to store an instruction in this logic section. This is done by writing the directive code with the section name before the instructions, such as: PROG1 code.

4.4.3 Library Manager MPLIB

A *library* is a collection of programs in a file, more commonly a collection of subroutines grouped over a common theme. These subroutines are available in the library file (file.lib). For example, it is possible to create a library through a group of subroutines related to mathematical operations and call it math.lib. The file with the library can be linked with the other files resulting from the assembling process as shown in figure 4.10. Using libraries has the advantage of having a single file with all the programs with a common focus. When the library is linked with the object files, the linker only takes from the library those subroutines that the program calls. This limits the size of the machine language program because it does not contain unnecessary code.

Libraries are created using the library manager MPLIB as shown in figure 4.11. This program can be invoked from the command line in the DOS operating system or from the integrated development environment MPLAB.

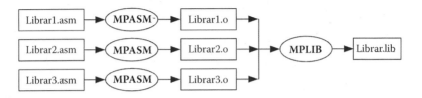

FIGURE 4.11
Process to obtain a library with MPLIB. Its different components are assembled with MPASM, obtaining the appropriate object files (.o). The library manager MPLAB creates the library file (.lib).

Example 4.42

Processes for creating and using a library. The library to be built (librar.lib) will consist of three basic subroutines called SR1, SR2, and SR3. Each subroutine has been programmed in independent files called librar1.asm, librar2.asm, and librar3.asm. These files are shown below. Pay special attention to the use of the global directive in these files.

File librar1.asm:

```
;
; Creation of a library.
; librar1.asm: this file contains subroutine SR1.
;
        list    p = 16f873
        #include p16f873.inc
        global  SR1
        code
SR1:
        nop
        return
        end
```

File librar2.asm:

```
;
; Creation of a library.
; librar2.asm: this file contains subroutine SR2.
;
        list    p = 16f873
        #include p16f873.inc
        global  SR2
        code
SR2:
        nop
        nop
        return
        end
```

File librar3.asm:

```
;
; Creation of a library.
; librar3.asm: this file contains subroutine SR3.
;
        list    p = 16f873
```

```
            #include p16f873.inc
            global   SR3
            code
SR3:
            nop
            nop
            nop
            return
            end
```

These three files are assembled using MPASM and are processed with the library manager MPLIB. The resulting file is library.lib.

The file program.asm contains the source code program. In this example, it is also a trivial program but it uses one of the subroutines created above. Note the use of the directive extern in the file.

```
; Program that uses one subroutine from the library library.lib.
; By examining the file program.lst it is seen that the linker only
uses the code
; that corresponds to the subroutine being called.
;
      list    p = 16f873
      #include p16f873.inc
      extern   SR2
Program  code 0
;
      call    SR2
;
      end
```

This program is assembled and linked with the file library.lib. The result of this process is the file program.hex that contains the object code as well as a list program (program.lst). It is possible to verify in this last file that the linker has only used the subroutine called. The following is a partial listing for the file program.lst:

```
Address Value    Disassembly   Source

                      ; Program that uses one subroutine from the
                        library library.lib.
                      ; By examining the file program.lst it is
                        seen that the linker only uses the code
                      ; that corresponds to the subroutine being
                        called.
                      list    p = 16f873
                      #include p16f873.inc
                              extern   SR2
                              Program code 0
                      ;
000000  2007    CALL  0x7    call    SR2
                      ;
                      end
                      ;
                      ; Creation of a library.
                      ; librar2.asm: this file contains subroutine
                        SR2.
                      ;
                      list    p = 16f873
```

```
                              #include p16f873.inc
                                      global   SR2
                                      code
                                      SR2:
000007  0000    NOP                   nop
000008  0000    NOP                   nop
000009  0008    RETURN                return
                                      end
```

5

Parallel Input and Output

This chapter describes the parallel input and output (I/O) resources in PIC microcontrollers. Parallel communication is a type of communication in which all the data bits are transferred simultaneously. Serial input and outputs are described in Chapter 8, and analog input and outputs are described in Chapter 9. This chapter starts by explaining the basic concepts and techniques associated with data transfer, followed by parallel ports in medium-end PIC microcontrollers. Finally, this chapter illustrates the connection of several peripherals widely used in microcontroller systems, such as switches, light-emitting diodes (LEDs), and keypads, as well as seven-segment displays and liquid-crystal displays (LCDs).

5.1 Basic Concepts

A *peripheral* is an external device connected to the microcontroller. The most widely used peripherals in microcontroller systems are switches, LEDs, relays, keypads, seven-segment displays and LCDs, A/D and D/A converters, printers, and motors. All these peripherals must include the required interface so they can be connected to a microcontroller port.

A *port* in a microcontroller is a circuit, internal to the microcontroller, used to interface it with peripherals or external devices. Figure 5.1 shows the general connection between a microcontroller and a peripheral using an I/O port. Generally, this connection has n lines (typically $n = 8$) to transfer data and m additional lines for data transfer control. These control lines may not be needed. For example, simple asynchronous I/O, described in further detail in Section 5.1.1, does not require control lines. It is important to notice that although the port shown in Figure 5.1 can transfer data in two directions, it is usual to find input-only or output-only ports.

From the programming point of view, ports are identified by their addresses that are usually part of the data memory. Therefore, at least one address is needed to represent the data entering or leaving the port. Control lines will require some additional bits in other addresses. In PIC microcontrollers, ports are accessed through the special function registers located in the data memory.

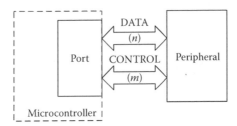

FIGURE 5.1
Connection between microcontroller and peripheral through an I/O port. The connection has n data lines and m control lines.

The basic element for any I/O port is a D latch, a device that can store 1 bit. An 8-bit parallel port has eight D latches. A latch has one data input (D), a control line (G), and a data output line (Q). When G = 0 the latch is blocked, holding the previous value. When G = 1, the latch outputs in Q the value that was in the input D. When G returns to 0, the latch holds in Q the value that was on the input D. In some cases it may be necessary for the data to be available in a tri-state device as shown in Figure 5.2. It is then necessary to add an additional control line OE# (output enable) to enable the digital output (DO) tri-state line. When OE# = 1, DO is kept in high impedance. When OE# = 0, then DO = Q.

In an input port, the inputs D come from the peripheral and the tri-state outputs are connected to the internal data bus in the microcontroller. For an output port, the connections are reversed. The following section describes the techniques for transferring data between the parallel port and the microcontroller.

5.1.1 Data Transfer Techniques

Data transfer between a peripheral and a port can be classified as:

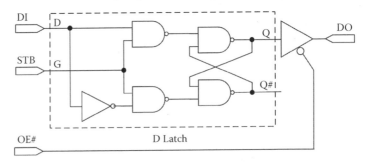

FIGURE 5.2
The basic element in an I/O port is the D latch used to store a bit. This figure shows the D latch with a tri-state data output. DI is the data input. The latch captures the data in DI with the signal STB. The signal OE# is used to enable the tri-state output DO.

FIGURE 5.3
Simple I/O for an 8-bit parallel data transfer. (a) Simple I/O without synchronization. (b) Simple I/O using STB signal. STB notifies when data is ready at the data pins. This notification can be done with voltage level (for example, STB = 1 as shown in the figure) or by its edges.

- Simple input/output
- Controlled input/output

Simple I/O is based on transferring the data bits between the port and peripheral without the use of any control signals, as shown in figure 5.3a. The connection of switches in the input lines or LEDs at the output lines in a parallel port are typical applications of simple I/O. Sometimes a synchronization signal (STB, strobe) is transmitted in addition to the data signals to indicate when the data is available. This indication can be by voltage levels or by the edge of the STB signal. For example, when using voltage levels, when the peripheral keeps the signal STB active (STB = 1) it means that the data is available in the port or peripheral data pins. The receiving device (either the port or the peripheral) must capture the data in sync with STB. Figure 5.3b shows this variation of the simple I/O technique. When the indication is done by signal edge, the data must be captured in sync with the appropriate edge of the signal STB.

In controlled I/O, there is a conversation, called *handshake*, between the port and the peripheral. Controlled I/O requires two or more control signals and a protocol that the port and peripheral must follow to understand each other. Figure 5.4 shows two variations of controlled I/O using two control signals. The control signal *strobe* (STB) is generated by the device that transmits the data and the control signal *acknowledgment* (ACK) is generated by the receiver.

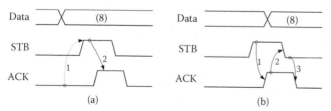

FIGURE 5.4
Controlled I/O for 8-bit data. (a) Simple I/O control. (b) Double I/O control.

In the first variation, shown in Figure 5.4a, the device transmitting the data sends the signal STB to the receiver indicating that the data is available at the data pins. In this case, the transmitter says to the receiver "I am sending you the data right now." The receiver captures the data and reports the action by activating the signal ACK (ACK = 1), saying to the transmitter "I have received the data and I am processing it." The transmitter does not send new data until the signal ACK has been set back to 0. By making ACK = 0, the receiver says to the transmitter "Send the next data." Therefore, the signal ACK works as an indication that the receiver is processing the data that has been received.

The second variation of controlled I/O uses a slightly more complicated protocol as shown in Figure 5.4b. First, the transmitter lets the receiver know that it will be sending data although this data may not be available yet. The transmitter does this activating the signal STB (STB = 1). Here, the transmitter says to the receiver "I am going to send you data. Can I do it?" Once the receiver detects the activation of the signal STB and is able to accept the data, then the receiver activates the signal ACK (ACK = 1). Now, the receiver says to the transmitter "Send it." The transmitter detects ACK = 1, indicating that it can now transmit the data. After the transmitter stores the data in the data pin, it indicates to the receiver that the data is now available by making STB = 0. With this, the transmitter says to the receiver "I am sending you the data." The receiver captures the data, processes it, and when it is able to accept new data will indicate this to the transmitter by making ACK = 0, that is, saying to the receiver "I have received your data. You can send new data."

The control signals and the logic conversation between the port and peripheral can be manipulated by hardware or by software. Hardware manipulation requires the port have circuits able to generate the signals STB and ACK without the intervention of the microcontroller. Software manipulation means that a program specifically created for this purpose generates the signals. PIC microcontrollers have parallel ports of up to 8 bits, all of them independent. They do not have specialized ports to implement hardware manipulation of the control signals; for this reason, controlled I/O must be implemented using software.

5.1.2 Input/Output Techniques

The two most commonly used techniques to service a peripheral connected to a microcontroller are:

- Programmed input/output
- Interrupt input/output

Programmed I/O is basically a software technique. It needs bits to indicate the status of the peripheral (ready or not ready). The program asks if the peripheral needs attention; if the response is affirmative, it carries out the appropriate action, which is normally the writing or reading of data in the port connected to the peripheral. If the response is negative, the program performs other tasks or simply waits until the peripheral is ready.

Figure 5.5 shows the algorithms used by the two main variants of programmed I/O: polling I/O and waiting I/O. In polling I/O, the microcontroller carries out other tasks if the peripheral is not ready to transmit or receive data. In waiting I/O, the microcontroller waits until the peripheral is ready. Obviously, polling I/O manages the time for the microcontroller better.

The main characteristic of the interrupt I/O technique is that the peripheral indicates the need for attention. This is done by sending an interrupt request to the microcontroller. When the microcontroller receives this signal, it interrupts the execution of the program and moves toward executing the interrupt subroutine. When the microcontroller finishes with the request from the peripheral, it continues with the program that had been interrupted. Interrupt I/O uses a combination of hardware and software. The hardware part is based on the circuits needed to request and execute the interrupt that will be studied in further detail in Chapter 7.

In both I/O techniques, the data transmission speed between peripheral and microcontroller is ultimately limited by the speed in executing instructions, because this execution is based on reading or writing data using the appropriate instructions. Furthermore, the data transferred between memory and the I/O ports normally have to move through the CPU, thus further limiting the speed of the process. Microprocessor systems utilize a third I/O technique to bypass these limitations by using direct memory

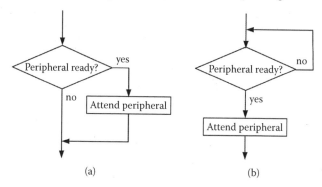

(a)   (b)

FIGURE 5.5
Variations of programmed I/O. (a) If peripheral is not ready, the program continues to work on other tasks. (b) If peripheral is not ready, the program waits until it is ready.

access (DMA). DMA is a hardware-implemented I/O technique based on the direct data transfer between peripheral and memory without having to execute a program, thus allowing for very high data-transfer rates. This technique is, however, used in very few microcontrollers. None of the PIC microcontrollers use DMA for I/O.

5.2 Parallel Ports in Medium-End PIC Microcontrollers

Medium-end PIC microcontrollers can have up to seven parallel ports named PORTA, PORTB, PORTC, ..., PORTG, with each of them having up to 8 bits. The pin ports are identified as RA$<x>$, RB$<x>$, ..., RG$<x>$ in which x is the number of the bit ($x = 0, 1, ..., 7$). Generally, each port line can be programmed as an input or as an output. Most of the pins in the I/O ports can carry out several functions. For example, the same pin can work as a digital input or output, or can be an analog input to the A/D converter, or can carry signals to or from the timers. Some PIC microcontrollers also have a parallel slave port (PSP) that behaves like an 8-bit generic bus with lines to control data transfer between the PIC and the peripheral. When a PIC microcontroller has a PSP it shares its pins with ports D and E.

Each parallel port has two special function registers used to manipulate the port. These registers are called PORT and TRIS (PORTA, TRISA, PORTB, TRISB, etc.). The PORT registers store the output data and the TRIS registers are used to program each line in the port as an input or as an output line.

Each bit in the TRIS register is programmed as:

```
TRIS<x> = 1 Programs port line <x> as an input.
TRIS<x> = 0 Programs port line <x> as an output.
```

Figure 5.6 shows the basic schematic for an I/O pin. The circuit contains two D latches: one of them is used to store the output data, while the other is used to store the control bit TRIS$<x>$. The I/O pin is manipulated by the two MOS transistors in totem-pole configuration. In this configuration, transistor T1 (P-channel transistor) is ON when the gate voltage is 0 and OFF when the gate voltage is 1. Transistor T2 (N-channel transistor) works in the opposite way. The totem pole is driven by an AND and an OR gate from the latches. From the circuit it is possible to realize that if the control latch stores a 1, both transistors are OFF. This causes the I/O port to be in a high-impedance state (third state), therefore acting like a data input pin. If the control latch stores a 0, the I/O port becomes an output pin. The value at the pin is equal to the value stored in the data latch. Table 5.1 shows the truth table for this circuit.

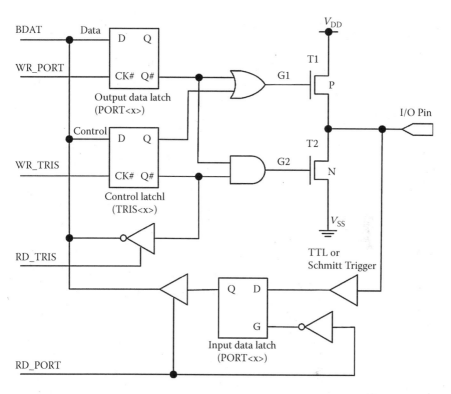

FIGURE 5.6
Basic schematic for a I/O pin in a medium-end PIC microcontroller.

As seen in Figure 5.6, this circuit uses an additional D latch that stores the state of the I/O pin when it has been configured as an input pin. The input D to this third latch is the voltage in the I/O pin connected through a noninverting gate that can be a TTL (transistor transistor logic) or Schmitt trigger. Writing data in a port means to write in the appropriate latch, whereas reading the bit in a port is equivalent to reading the logic state (voltage value) in the pin. This means that if data is written in an output port and this port is read, the value may be different. This can happen especially if the maximum value for output currents is exceeded.

TABLE 5.1
Truth Table for the Circuit in Figure 5.6

Control	Data	G1	G2	T1	T2	I/O Pin
1	x	1	0	OFF	OFF	Input: Hi Z
0	0	0	0	OFF	ON	Output: V_{SS} (0)
	1	1	1	ON	OFF	Output: V_{DD} (1)

When connecting devices to the port pins it is necessary to take into account the power limitations for the microcontroller. Each I/O pin can source or sink a maximum value of current. Furthermore, the total current sourced or sunk by all the port pins cannot exceed a certain value. This value is normally lower than the sum of all the allowed individual currents for each pin in the port. Finally, to keep the voltage in the output pin within the limits for the logic values 0 and 1, it is necessary to keep the output currents for high level (I_{OH}) and low level (I_{OL}) within the limits specified by the manufacturer. All the devices must comply with these specifications described in the manufacturer's data sheets.

Example 5.1

The following voltage and current values for the PIC16F873 pins are given by its manufacturer. Each I/O design must take them into consideration.

1. Maximum sunk current by any pin: 25 mA.
2. Maximum sourced current by any pin: –25 mA.
3. Sourced or sunk current for port C or by ports A and B together cannot exceed 200 mA.

Typical values for input and output voltages:

1. Logic value 0 in an input pin: Voltage in this pin is $V_{IL} < 0.75$ V (TTL input) or $V_{IL} < 1.0$ V (Schmitt trigger input) with $V_{DD} = 5.0$ V.
2. Logic value 1 in an input pin: Voltage in this pins is $V_{IH} > 2.0$ V (TTL input) or $V_{IH} > 4.0$ V (Schmitt trigger input) with $V_{DD} = 5.0$ V.
3. Output voltage for logic value 0: $V_{OL} < 0.6$ V if output current sunk by pin is $I_{OL} < 8.5$ mA with $V_{DD} = 4.5$ V.
4. Output voltage for logic value 1: $V_{OH} > V_{DD} - 0.7$ V $= 3.8$ V if output current sourced by pin is $I_{OH} < -3.0$ mA with $V_{DD} = 4.5$ V.

The selective or individual manipulation of bits in a port requires some caution. In reality, PIC microcontrollers do not have the necessary hardware in their ports to exclusively manipulate an output bit. Therefore, to modify a single bit, all the bits in the port need to be written. Even the specific bit manipulation instructions, such as bcf and bsf used to set any bit to 0 or 1, operate by reading the register, modifying the bit specified in the instruction, and writing the resulting word back in the register. That is, the selective modification of a bit in a register is an operation that involves reading, modification, and writing the complete register. When modifying a bit in the data register, this operation is transparent for the programmer. However, when the bit being modified is a bit in the PORT register of a parallel port, the manner in which this operation is carried out may produce unexpected results in the other port bits. This is because when a port is read, the value that is actually read is the status of its pins and not the value stored in the port. Therefore, the resulting value may be different from the real value in the PORT register.

5.2.1 Port A

Port A can have up to 8 bits although most of the medium-end PIC micro-controllers (such as the PIC16F873) have only 6 bits implemented. These bits correspond to pins RA0 to RA5. All these pins can be configured as input or output terminals. RA4 is a Schmitt trigger input and when pro-grammed as an output it becomes an open-drain output. The special func-tion registers associated with port A are PORTA and TRISA.

Port A pins may be shared with the inputs for the A/D converter if the microcontroller has one, such as in the case for the PIC16F873. In this case, port A pins can be digital or analog. This is programmed with the spe-cial function register ADCON1. The pin RA4 is also used as the external clock input for the timer Timer0. In this case, the pin is called RA4/T0CK1. Table 5.2 shows the functions of the port A pins in a PIC16F873.

Example 5.2

Port A pins are also analog inputs for those PICs that incorporate an internal A/D converter. This example shows how to program port A for microcontrollers without an internal A/D converter as well as those with it.

Initialize port A (PIC without internal A/D converter):

```
clrf     STATUS        ; Select bank 0.
clrf     PORTA         ; Set PORTA register to 0.
bsf      STATUS, RP0   ; Select bank 1.
movlw    0xCF          ; Value in TRISA to program
movwf    TRISA         ; RA<3:0> as inputs and RA<5:4>
                       ; as outputs.
```

TABLE 5.2

Pin Functions in PIC16F873

Name	Function
RA0/AN0	Digital input/output or analog input
RA1/AN1	Digital input/output or analog input
RA2/AN2	Digital input/output or analog input
RA3/AN3/VREF	Digital input/output or analog input, or reference voltage for A/D converter
RA4/T0CKI	Digital input/output or external clock input for Timer0. Open-drain output
RA5/SS/AN4	Digital input/output or input for selection synchronous serial port or analog input

```
        bcf        STATUS, RP0    ; Select bank 0.
```

Initialize port A (PIC with internal A/D converter):

```
        bcf        STATUS, RP0    ; Select bank 0.
        bcf        STATUS, RP1    ;
        clrf       PORTA          ; Set PORTA register to 0.
        bsf        STATUS, RP0    ; Select bank 1.
        movlw      0x06           ; Configure all port terminals
        movwf      ADCON1         ; as digital input or outputs.
        movlw      0xCF           ; Value in TRISA to program
        movwf      TRISA          ; RA<3:0> as inputs and RA<5:4>
                                  ; as outputs.
        bcf        STATUS, RP0    ; Select bank 0.
```

5.2.2 Port B

Port B has 8 bits with pins called RB0 to RB7. All these pins can be configured as input or output pins using the special function register TRISB. The special function register PORTB is used to write data in port B. Table 5.3 shows the function of the port B pins in a PIC16F873. Each port B pin has an internal pull-up circuit that can be programmed with the bit RBPU# in the special function register OPTION (bit OPTION <7>). This bit enables or disables the pull-up in port B.

TABLE 5.3

Port B Pins in PIC16F873

Name	Function
RB0/INT	Digital input/output; external interrupt input
RB1	Digital input/output
RB2	Digital input/output
RB3/PGM	Digital input/output; in-circuit programming pin
RB4	Input/output. As input pin, interrupts may be programmed by a change in logic level.
RB5	Input/output. As input pin, interrupts may be programmed by a change in logic level.
RB6/PGC	Input/output. As input pin, interrupts may be programmed by a change in logic level. In-circuit programming pin.
RB7/PGD	Input/output. As input pin, interrupts may be programmed by a change in logic level. In-circuit programming pin.

Note: All pins have a software-programmable internal pull-up circuit. RB0 can be used as an external interrupt input. A voltage change in inputs RB4 to RB7 can generate an interrupt request. Pins RB3, RB6, and RB7 are used for in-circuit programming. This allows for programming the microcontroller in the same board that will be used to run the intended application.

An important aspect of port B is that it can generate an interrupt request by changing the logic level in any of the pins RB4 to RB7. If these pins are programmed as inputs, a change in the input logic level from 0 to 1 or from 1 to 0 generates an interrupt. This change can be produced, for example, by pressing a key connected to one of the port pins. When this type of interrupt is produced, bit RBIF in the INTCON register (bit INTCON <0>) is set at 1. The bit RBIE in the INTCON register (bit INTCON<3>) is used to enable or disable interrupts. This interrupt can be used to wake up the microcontroller from a low-power mode.

Pin RB0 can also accept an edge-triggered external interrupt request. In this case, the pin is called RB0/INT. This interrupt is reported in bit INTF in the INTCON register (bit INTCON <1>) and is enabled or disabled by bit INTE in the INTCON register (bit INTCON <4>). The bit INTEDG in the OPTION register (bit OPTION <6>) selects the interrupt for the raising or falling edge.

Pins RB3, RB6, and RB7 in port B can be used for In-Circuit Serial Programming (ICSP). This is a resource in PIC microcontrollers that allow programming of the device on the same board that will be used to execute the intended application. The program is sent to the OTP, EEPROM, or flash memory using a serial transmission format through these pins. See the device programming specification before using this resource.

5.2.3 Port C

Port C is an 8-bit parallel port with pin names RC0 to RC7. Writing in port C is done by using the special function register PORTC. All pins can be configured as Schmitt trigger inputs or digital outputs using the special function register TRISC. Pins in port C share functions with other input and output devices: Timer1, the Compare/Capture/PWM (CCP) module, and the Synchronous Serial Port (SSP) or Master Synchronous Serial Port (MSSP), and Universal Synchronous Asynchronous Transmitter Receiver (USART). These functions are shown in table 5.4 for the PIC16F783.

5.2.4 Ports D, E, F, and G

Ports D, E, F, and G are parallel ports of up to 8 bits. All pins can be programmed as digital inputs or outputs. When configured as inputs, they are Schmitt trigger inputs. When these ports exist, their functions are shared with the functions from the PSP. Some of the pins in port E can also be used as analog inputs in addition to those existing in port A. The special function registers associated with ports D and E are PORTD and PORTE for data, and TRISD and TRISE for control. The PIC16F874 has ports D and E, but the PIC16F873 does not have them.

Ports F and G are parallel ports of up to 8 bits with Schmitt trigger inputs. Their function is shared with the outputs for LCD drivers. Ports

TABLE 5.4

Port C Pins in the PIC16F873

Name	Function
RC0/T1OSO/T1CKI	Digital input/output or output for Timer 1 or clock input for Timer 1
RC1/T1OSI/CCP2	Digital input/output or input for Timer 1 or clock input for Timer 1 or pin for CCP2 module
RC2/CCP1	Digital input/output or pin for module CCP1
RC3/SCK/SCL	Digital input/output or pin for synchronous serial port
RC4/SDI/SDA	Digital input/output or pin for synchronous serial port
RC5/SDO	Digital input/output or pin for synchronous serial port
RC6/TX/CK	Digital input/output of pin for USART serial port
RC7/RX/DT	Digital input/output of pin for USART serial port

Note: All inputs are Schmitt trigger inputs.

F and G only exist in those microcontrollers specifically manufactured to drive LCDs directly.

5.2.5 Parallel Slave Port (PSP)

The PSP is an 8-bit directional port that carries control signals to read and write data from an external device. The PSP can be used to connect the microcontroller directly to the data and control bus in a microprocessor- or microcontroller-based system. This makes the PIC with the PSP become an I/O port of that system as shown in Figure 5.7a.

For those microcontrollers that incorporate a PSP, such as the PIC16F874, the PSP is implemented in the pins for ports D and E. PSP has eight data lines (PSP<0:7>) and three control lines for reading data (RD#), writing data (WR#), and selection (CS#). The PSP data lines are implemented on the pins RD<0:7> and the control lines are implemented on the pins RE<0:2>. When the system external to the PIC wants to write or read data from the PIC using the PSP, it must select the device with CS# = 0 during data writing or reading. If CS# = 1, the data lines are kept at high impedance. During the reading process, the data pins act as outputs; during the writing process the data pins act as inputs. Figure 5.7b shows the PSP signals during the writing and reading cycles.

Three bits in the microcontroller inform of the state of the PSP. These are bits IBF and OVF in the TRISE register and bit PSPIF in the register PIR1. The input buffer full (IBF) bit is set to 1 when the PORTD register holds data written in the PSP from outside; IBF is set automatically to 0 when the program reads the input data in PORTD. The output buffer

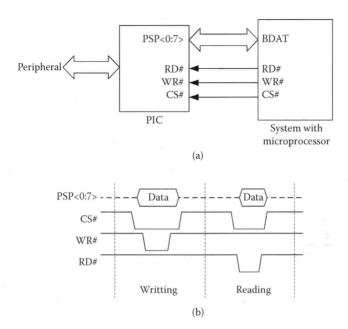

FIGURE 5.7
The parallel slave port (PSP) is an 8-bit bidirectional bus with control signals to transfer data between the PIC and an external device. (a) Potential use of a PSP to directly connect the PIC to a data bus in a system with a microprocessor or microcontroller. The PIC becomes an I/O port of this system. (b) Signals involved in I/O data transfer using the PSP.

full (OBF) bit is set to 1 when the program writes data in the PORTD register. This data must leave the microcontroller through the PSP pins. OBF is set to 0 when the external device has read that output data in the PORTD register.

The PSP interrupt flag (PSPIF) bit is set to 1 each time data is read or written from the exterior. It must be reset to 0 by software once the program has serviced the PSP transfer. PSPIF = 1 generates an interrupt request if the PSP has its interrupts enabled. Setting the PSP interrupt enabled (PSPIE) bit to 1 enables these interrupts.

The management of data input/output through the PSP can be either programmed or performed by interrupts. Programmed I/O is carried out by checking bits PSPIF, IBF, and OBF. In interrupt I/O, the program that handles the PSP interrupt must check bits IBF and OBF to find out the type of transfer that happened between the PSP and the exterior.

5.3 Connection of Commonly Used Peripherals

5.3.1 Switches and LEDs

Switches and LEDs are I/O devices commonly used in systems with micro-controllers. Figure 5.8 shows three possible ways for connecting these devices to the pins in port B of a PIC microcontroller.

LED1 is ON when the pin RB<i>, which should have been configured as a digital output, is low. In this configuration the LED1 current I_1 enters into the port pin. To keep the 0 logic level when LED1 is ON, it is necessary that

$$I_1 \leq I_{OLmax} \tag{5.1}$$

with I_{OLmax} being the maximum output current for the pin to stay at a logic low level.

LED2 is ON when with the pin RB<j> is high. In this configuration the LED2 current leaves the pin. For the pin to maintain its high logic level, it is necessary that

$$I_2 \leq I_{OHmax} \tag{5.2}$$

with I_{OHmax} being the maximum output current for the pin to stay at a high level.

In general, $|I_{OLmax}| > |I_{OHmax}|$, making condition 5.1 easier to meet than condition 5.2. For this reason, when connecting LEDs to the parallel ports it is recommended to use the configuration as shown for LED1.

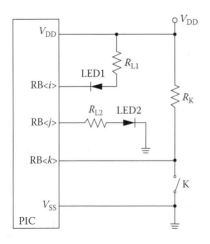

FIGURE 5.8
Connection of LEDs and switches to port B in a PIC microcontroller.

Example 5.3

Using a PIC16F873, connect 2 LEDs and a switch as shown in Figure 5.8. Calculate the values of resistances R_{L1}, R_{L2} assuming the voltage supply is $V_{DD} = 5$ V.

The LEDs most commonly used as indicators have forward current $I_F = 10$ mA and forward voltage about 2.0 V. These values change slightly depending on the LED color.

With $I_F = 10$ mA and $V_{RB<i>} = V_{OL}$ ($V_{OL} = 0.35$ V for $I_{OL} = 10$ mA, from PIC data sheet), the value for R_{L1} for a red LED ($V_F = 1.6$ V), is

$$R_{L1} = \frac{V_{DD} - V_F - V_{OL}}{I_F} = \frac{5 - 1.6 - 0.35}{0.01} = 305\,\Omega.$$

With $I_F = 10$ mA and $V_{RB<j>} = V_{OH}$ ($V_{OH} = 4.3$ V for $I_{OH} = 10$ mA, from the PIC data sheet), the value for R_{L2} is

$$R_{L2} = \frac{V_{OH} - V_F}{I_F} = \frac{4.3 - 1.6}{0.01} = 270\,\Omega.$$

The value for the resistance in series with the LED is not critical. Therefore it is possible to use standard values with tolerances of 5% or even 10%. This example could use $R_{L1} = 300\,\Omega$ and $R_{L2} = 270\,\Omega$, with 5% tolerance.

Switches are connected to digital inputs. Switch K in Figure 5.8 is connected to the input RB<k>. In this configuration, when the switch is closed, the voltage at the pin is 0 V (low level). When the switch is open, the pull-up resistance R_K guarantees a high level at the pin input. The value of the pull-up resistance can be on the order of tens of kiloohms because the input current to the microcontroller is very low. Pins in port B have an internal pull-up that can be connected or disconnected using the bit RBPU in the OPTION register. If the switches are connected to pins in terminal B and the internal pull-ups have been connected, it is not necessary to use the external resistors R_K.

Mechanical switches are essentially two metallic pieces that come in contact with each other. These mechanical switches are affected by a problem known as bouncing. When a switch is being open or closed, the metallic pieces do not reach their final position immediately; instead they act like a ball being dropped on the floor—they bounce for a certain period of time. This originates fast changes in the contact resistance before the switch reaches its steady state as shown in Figure 5.9. This bouncing may cause a single switch action, either opening or closing, to be understood by the microcontroller as a series of successive switch activations and deactivations.

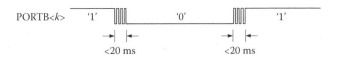

FIGURE 5.9
Bouncing in mechanical switches. When the switch opens or closes, the mechanical pieces vibrate generating fast contacts before settling. This transient normally lasts less than 20 ms.

Bouncing can be solved by using hardware or software methods. A basic hardware solution is to use a nonmechanical switch such as Hall-effect switches or wet-contact (mercury) switches that are not affected by bouncing. However, the number of available models for this type of switch is very limited compared to the traditional mechanical models. A better hardware solution is to connect the mechanical switch to the input of a monostable circuit that will increase the time for the first detected pulse long enough to mask the pulses produced by bouncing. This solution, however, increases the size and cost of the design.

Software solutions are based on reading the state of the switch a certain amount of time after the switch was first activated. A delay of about 20 ms is normally long enough to ensure that the switch has reached a stable state. Example 5.4 shows how to implement this solution.

Example 5.4

This example shows how to read the state of a switch connected to bit k in port B as shown in the circuit in figure 5.8. The problem of switch bouncing is solved by software using the algorithm shown in figure 5.10.

In this algorithm, the state of the switch is first read and stored in a register (TEMP) in the microcontroller. After waiting 20 ms to ensure there are no bouncing effects, the state of the switch is read again. If the values obtained in both cases are the same, then the readings are validated having ensured the correct reading of the state of the switch. However, if the values are different, it means that there are still bouncing effects. In this case, the microcontroller continues reading the state of the switch at 20 ms intervals until the two values are equal.

The subroutine READ_K shown below implements the algorithm shown in figure 5.10.

```
        list p = 16f873
        #include <p16f873.inc>
TEMP  equ   0x20     ; Register for intermediate data storage.
k     equ   3        ; Number of bit in port B to which the
                     ; switch is connected.
; READ_K: Subroutine to read the read the state of a switch
; connected to port B
;     avoiding bouncing effects
;     Inputs: None
;     Outputs: In W<0>, the value of bit PORTB<k>
```

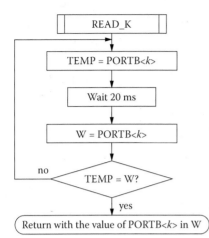

FIGURE 5.10
Algorithm used to read a switch solving the problem of bouncing. The switch is connected to bit *k* in port B as shown in Figure 5.8. TEMP is a data memory register and W is the working register.

```
READ_K:
    btfss    PORTB, k ; Read port B. Is bit PORTB<k> = 1?
    goto     K0       ; No: store 0 in TEMP.
    movlw    1        ; Yes: store 1 en TEMP.
    movwf    TEMP     ; TEMP to 1.
    goto     K1
K0:
    clrf     TEMP     ; TEMP to 0.
K1:                   ; Value of PORT<k> is in TEMP.
                      ;
    call     DEM20    ; Wait 20 ms.
                      ;
    btfss    PORTB,k  ; Read again port B. PORTB<k> = 1?
    goto     K2       ; No: Store 0 in W.
    movlw    1        ; Yes: Store 1 in W.
    goto     K3
K2:
    clrw              ; W in 0.
K3:                   ; Value of PORT<k> is in W.
    xorwf    TEMP,W   ; Compare the two read values. If they are equal,
                      ; store
                      ; 0 in W. Z is activated. TEMP does not change.
    btfss    STATUS,Z ; TEMP = W? Z = 1?
    goto     READ_K   ; No: Switch still bouncing. Read PORTB<k> again.
K4:                   ; Yes: Bouncing ended. Finish subroutine.
    movf     TEMP, W  ; Store value of PORTB<k> in W and
    return            ; return.
; DEM20: Routine to wait for 20 ms.
DEM20
    ;
    ; Write here the code for this subroutine.
    ;
    return
    end
```

5.3.2 Matrix Keypads

Matrix keypads consist of keys interconnected in the shape of a matrix. Each key is a simple mechanical switch located at the crossing between the matrix rows and columns. When a key is pressed, its row and column form an electrical contact. Row and columns can be connected to the pins of one or more parallel ports. Figure 5.11 shows a 16-key matrix keypad arranged in four rows and four columns.

The state for a matrix keypad can be explored by sending signals through its rows (exploration lines) and reading the information received through its columns (return lines). When none of the keys are pressed, all the return lines will have the logic state 1. The exploration lines are then set to 0 either simultaneously or sequentially. Only the return line that links the pressed key with its exploration line will read a low logic value. The rest of the return lines will be read as 1. The information sent to and received from the matrix makes up a code unique for each key, known as an exploration code. To ensure that the return lines are kept at 1 when no keys are pressed, it is necessary to connect resistors between each return line and the voltage supply V_{DD} as shown in figure 5.11.

The procedure to service matrix keypads is as follows:

Step 1: Wait until the keypad is clear (due to an earlier key pressed).

Step 2: Detect that a new key has been pressed.

Step 3: Explore the matrix keypad to determine the key that was pressed. The exploration code, containing the row and column numbers is generated in this step. The exploration code can be generated in two ways:

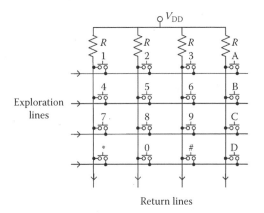

FIGURE 5.11
A 16-key matrix keypad.

- Sequential exploration of rows. This method sets the first row to 0 and reads all the columns. If none of the columns is read at 0 it means that the pressed key is not in that row. Then the next row is set to 0, and the columns are read again. This process is repeated until a 0 is found in a column. This determines the row and column for the pressed key, thus giving the exploration code for that key.

- Simultaneous exploration of rows and columns. This method sets all the rows to 0 and reads all the columns. This detects the column that contains the pressed key but not its row. Then the process is inverted: all the columns are set to 0 and the rows are read. This detects the row that contains the pressed key. This gives the row and column for the pressed key, and thus the exploration code for the key.

The simultaneous exploration method allows exploring the keypad faster, in only two steps. However, it requires the exploration and return lines to be bidirectional (although this is not a problem in PIC microcontrollers, it can be a problem in other devices). The sequential exploration method is slower but allows for exploration and return lines to be unidirectional.

Figure 5.12 shows the algorithm for the sequential exploration method. This basic algorithm could be improved by adding the necessary steps to check for the validity of the read code, as well as to determine if more than one key had been pressed simultaneously. When using an alphanumeric keypad, for example, the algorithm could be further improved by converting the exploration code into the ASCII code for the pressed key.

Example 5.5

Subroutine for servicing a 16-key matrix keypad connected to port B of a medium-end PIC microcontroller.

Figure 5.13 shows a possible method for connecting the matrix keypad to port B. When comparing this approach to the schematic shown in figure 5.11, there are two main differences. First, the return lines do not have pull-up resistors because the pins in port B have internal pull-ups that guarantee a 1 in the return lines with no key being pressed. Second, there are four diodes (D) in the exploration lines. These diodes are used to limit the current in pins RB0 to RB3 if two keys in the same column were pressed at the same time. These diodes could be substituted by resistors between 1 kΩ and 2.2 kΩ.

The subroutine READKEY explores the matrix keypad following the algorithm shown in figure 5.12. This subroutine returns the exploration code for the pressed key. Table 5.5 shows the exploration codes.

The code for the subroutines INITKEY and READKEY are shown below. The subroutine INITKEY is used to prepare port B to service the keypad.

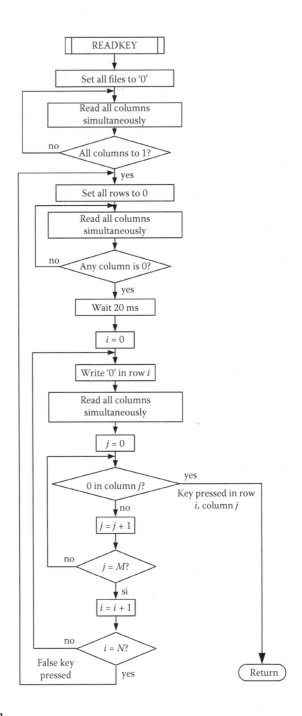

FIGURE 5.12

Algorithm to read an *N*-row, *M*-column matrix keypad using sequential exploration. The algorithm waits until a key is pressed and returns its position (row *i*, column *j*).

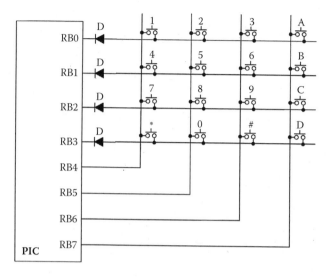

FIGURE 5.13

A 16-key matrix keypad connected to port B in a PIC. Diodes are used to avoid short circuits between two exploration lines if two or more keys in the same column are pressed simultaneously. This configuration uses the internal pull-up (not shown) available in port B to keep a logic 1 in the return lines when no keys are pressed.

TABLE 5.5

Exploration Codes Returned by the Subroutine READKEY When Exploring the Keypad Shown in Figure 5.13

Key	Row (Binary)	Column (Binary)	Exploration Code (Hexadecimal)
1	00	00	00
2	00	01	01
3	00	10	02
4	01	00	04
5	01	01	05
6	01	10	06
7	10	00	08
8	10	01	09
9	10	10	0A
0	11	01	0D
*	11	00	0C
#	11	10	0E
A	00	11	03
B	01	11	07
C	10	11	0B
D	11	11	0F

```
        list p = 16f873
        #include <p16f873.inc>
; Declarations:
    TEMP      equ    0x20          ; Temporary register used by
                                   ; subroutines.
    ROW       equ    0x21          ; Temporary register used by
                                   ; subroutines.
    COLUMN    equ    0x22          ; Temporary register used by
                                   ; subroutines.
    ; INITKEY: Subroutine to program port B.
    INITKEY:
            clrf      STATUS       ; Select bank 0.
            bcf       INTCON,INTE  ; Disable external interrupt by
                                   ; RB0.
            bcf       INTCON, RBIE ; Disable interrupt by changes in
                                   ; RB<7:4>.
            movlw     0FFh         ; Store value in PORTB.
            movwf     PORTB        ; Set all outputs in port B to '1'.
            bsf       STATUS, RP0  ; Select bank 1.
            movlw     0F0h         ; Load this value in TRISB
                                   ; to program
            movwf     TRISB        ; RB<3:0> as outputs and RB<7:4> as
                                   ; inputs.
            bcf       OPTION_REG,  ; Enable internal pull-ups in port
                      NOT_RBPU     ; B.
            bcf       STATUS, RP0  ; Select bank 0.
            return
    ; READKEY: Subroutine for exploring matrix keypad.
    ; This subroutine waits for a key being pressed. It returns its
    ; exploration code in W.
    ; Inputs: none
    ; Outputs: Exploration code in W. Bits W<3:2> contain the
    ; column for the pressed key. Bits W<1:0> contain the row.
    ;
    READKEY:
            movlw     0F0h
            movwf     PORTB        ; Set all rows to '0'.
            nop
    KEY10:                         ; Waits for keypad being free:
            movf      PORTB, W     ; Read all columns simultaneously.
                                   ; Low part of W
                                   ; is 0. High part of W, if a key is
                                   ; already pressed the bit
                                   ; that corresponds to the return
                                   ; line for the key
                                   ; is 0. The other bits are 1.
            xorlw     0F0h         ; Reverse situation. High part of W:
                                   ; Bit
                                   ; corresponding to returning line in
                                   ; pressed key
                                   ; is 1. Other bits are 0. Low part
                                   ; of W
                                   ; does not change. Z becomes active
                                   ; if no key is pressed.
            btfss     STATUS, Z    ; All keys open? Z = 1?
            goto      KEY10        ; No - wait for keypad being free.
```

```
KEY20:                              ; Yes - continue. Wait for a key
                                    ; being pressed.
        movlw   0F0h
        movwf   PORTB               ; All rows to '0'.
        nop
        movf    PORTB, W            ; Read all columns simultaneously.
        xorlw   0F0h                ; Z = 0 if a key is pressed.
        btfsc   STATUS, Z           ; Any pressed key? Z = 0?
        goto    KEY20               ; No - Wait for pressed key.
KEY30:                              ; Yes - continue.
        call    DELAY20             ; 20 ms delay for debouncing.
KEY40:                              ; Explore keypad rows to find out
        movlw   0FFh                ; what is the pressed key.
        movwf   PORTB               ; All rows to '1'.
        nop
ROW0:                               ; Explore row 0:
        movlw   0
        movwf   ROW
        bcf     PORTB, 0            ; Set row 0 to 0 (RB0).
        nop
        movf    PORTB, W            ; Read all columns simultaneously.
        call    IDENTIFY            ; If a key is pressed, identify
                                    ; column
                                    ; for key.
        btfsc   STATUS, C           ; Pressed key? C = 1?
        goto    KEY50               ; Yes - Found pressed key. Finish
                                    ; exploration.
        bsf     PORTB, 0            ; No - Set explored row to 1 and
                                    ; move to next.
        nop
ROW1:                               ; Explore row 1:
        movlw   1
        movwf   ROW
        bcf     PORTB, 1            ; Set row 1 to 0(RB1).
        nop
        movf    PORTB, W            ; Read all columns simultaneously
        call    IDENTIFY            ; If a key is pressed, identify
                                    ; column
                                    ; for key.
        btfsc   STATUS, C           ; Pressed key? C = 1?
        goto    KEY50               ; Yes - Found pressed key. Finish
                                    ; exploration
        bsf     PORTB, 1            ; No - Set explored row to 1 and
                                    ; move to next.
        nop
ROW2:                               ; Explore row 2:
        movlw   2
        movwf   ROW
        bcf     PORTB, 2            ; Set row 2 to 0 (RB2).
        nop
        movf    PORTB, W            ; Read all columns simultaneously
        call    IDENTIFY            ; If a key is pressed, identify
                                    ; column
                                    ; for key.
        btfsc   STATUS, C           ; Pressed key? C = 1?
        goto    KEY50               ; Yes - Found pressed key. Finish
                                    ; exploration
        bsf     PORTB, 2            ; No - Set explored row to 1 and
                                    ; move to next.
        nop
ROW3:                               ; Explore row 3:
        movlw   3
```

```
        movwf   ROW
        bcf     PORTB, 3        ; Set row 3 to 0 (RB3).
        nop
        movf    PORTB, W        ; Read all columns simultaneously
        call    IDENTIFY        ; If a key is pressed, identify
                                ; column
                                ; for key.
        btfsc   STATUS, C       ; Pressed key? C = 1?
        goto    KEY50           ; Yes - Found pressed key. Finish
                                ; exploration
        bsf     PORTB, 3        ; No - Set explored row to 1 and
                                ; move to next.
        Nop                     ; If arriving here withoutfinding a
                                ; pressed key
                                ; means false key pressed. Wait for
                                ; new key pressed
        goto    KEY20           ; Go to wait for a key beingpressed.
KEY50:                          ; Found a pressed key. Build
                                ; exploration code with
                                ; 4 bits: ROW: COLUMN.
        rlf     ROW, f          ; Row number in
        rlf     ROW, W          ; bits 2 and 3
        andlw   0FCh            ; in W register.
        iorwf   COLUMN, W       ; Column number in bits 0 and 1 in W.
        andlw   0Fh             ; Set high part of W to 0.
        return                  ; Return exploration code in W<3:0>.
; IDENTIFY: Subroutine to identify the column with the pressed key.
; Inputs: In W, reading of port B
; Outputs: Bit C in STATUS set to 1 if there is a pressed key.
;       Otherwise, C set to 0.
;       Number of column for pressed key stored in
;       register named COLUMN.
;
IDENTIFY:
        movwf   TEMP            ; Input information stored in
                                ; register TEMP.
        btfsc   TEMP, 4         ; Is bit of column 0 equal to 0?
        goto    COL1            ; No - Move to examine next column.
        movlw   0               ; Yes - There is a pressed key in
                                ; that column.
        movwf   COLUMN          ; Store number of column in register
                                ; COLUMN.
        goto    IDENT_FIN       ; Finish subroutine.
COL1:
        btfsc   TEMP, 5         ; Is bit of column 1 equal to 0?
        goto    COL2            ; No - Move to examine next column.
        movlw   1               ; Yes - There is a pressed key in
                                ; that column.
        movwf   COLUMN          ; Store number of column in register
                                ; COLUMN.
        goto    IDENT_FIN       ; Finish subroutine.
COL2:
        btfsc   TEMP, 6         ; Is bit of column 2 equal to 0?
        goto    COL3            ; No - Move to examine next column.
        movlw   2               ; Yes - There is a pressed key in
                                ; that column.
        movwf   COLUMN          ; Store number of column in register
                                ; COLUMN.
        goto    IDENT_FIN       ; Finish subroutine.
COL3:
        btfsc   TEMP, 7         ; Is bit of column 3 equal to 0?
```

```
            goto    COL4        ; No - Move to examine next column.
            movlw 3             ; Yes - There is a pressed key in
                                ; that column.
            movwf   COLUMN      ; Store number of column in register
                                ; COLUMN.
            goto    IDENT_FIN   ; Finish subroutine.
COL4:
            bcf     STATUS, C   ; Indicates key not pressed with
                                ; C = 0.
            return              ; Return.
IDENT_FIN:
            bsf     STATUS, C   ; Indicates key pressed with C = 1.
            return              ; Return.
; DELAY20: Subroutine to introduce a 20 ms delay.
DELAY20
            ;
            ; Write here code for the subroutine that introduces a 20
              ms delay
            ;
            return
            end
```

5.3.3 Seven-Segment LEDs

Seven-segment LEDs are mainly used to represent numeric information. Figure 5.14 shows their internal circuitry as well as its symbols. Seven-segment LED displays can be configured as common anode or as common cathode. In common anode connection, in order for a segment to light up, its connection pin must be driven with a low voltage (logic value 0 for positive logic), with the common anode being set to a high positive voltage (V_{DD}). In common cathode connections this situation is reversed:

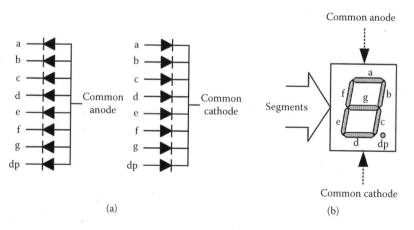

FIGURE 5.14

Seven-segment displays consist of a set of LEDs connected with a common anode or a common cathode. (a) The internal circuitry. (b) The seven-segment display symbol.

each segment is activated by a high voltage that corresponds to the logic level 1 while the common cathode must be set to 0 V (V_{SS}).

When using several seven-segment units making a display, their connections to the microcontroller parallel ports should be multiplexed. Figure 5.15 shows a connection schematic for several common anode seven-segment displays. All the similar segments have been connected to one another and to the pins in port B through current limiting resistors R_S. Each segment becomes active when its pin in the microcontroller has a logic value of 0. The common anode for each device acts as a selection line that is controlled from port A. Each seven-segment display is activated sequentially using an appropriate refresh frequency. In sync with

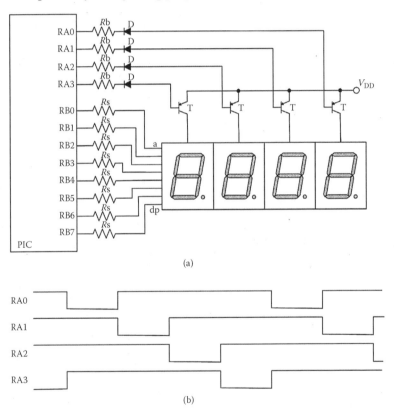

(a)

(b)

FIGURE 5.15

(a) Circuit for four common anode seven-segment displays connected to ports A and B in a PIC microcontroller. Resistors R_s limit the current through the segments, and resistors R_b ensure the saturations of the transistors used to enable each device. The diodes ensure transistors will be OFF when the pins in port A are at the logic level 1. (b) Waveforms at each selection pin. The frequency of these signals must range between 40 Hz and 200 Hz for the human eye not to detect blinking. Each device is selected one-fourth of the total time.

its activation, the information corresponding to the selected device is set in each pin. Using a high enough refreshing frequency (between 40 and 200 Hz), a person will see all the seven-segment displays lighted at the same time.

In a multiplexed display, each digit is ON for only a fraction of the total time (for example, one-fourth of the total time in the example shown in Figure 5.15). Therefore, it is necessary to increase the current in each segment in approximately the same proportion (×4) to achieve a good luminance level. If, for example, the forward current for a specific LED is 10 mA, it will be necessary to drive it with 40 mA pulses when working multiplexed. This current value can be adjusted with the resistance R_s. Resistance R_b must guarantee that the transistors will saturate. The diodes placed in the transistor bases ensure that these will be OFF when not directly driven.

It is possible to use an interrupt to handle the circuit in Figure 5.15. Figure 5.16 shows the block diagram for the algorithm to service the interrupt in that circuit. When multiplexing four devices it is necessary to use an oscillator with a frequency four times higher than the desired refreshing frequency to generate the interrupt signals to the microcontroller. Each interrupt is used to visualize one display. It is possible to use four registers in the data memory to store the "image" for the information that needs to be displayed. This is reflected in Table 1 in Figure 5.16. Each position in this table stores the seven-segment code for the desired display. Four additional registers (Table 2) can store the control information needed to select the appropriate devices. These are the words that will be sent to port A to correctly select the displays sequentially. Finally, one last register can be used to store a digit pointer that must be refreshed by the interrupt attention program. During each interrupt,

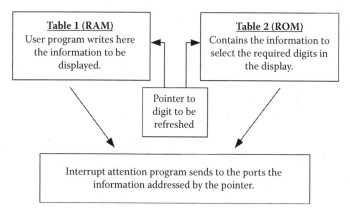

FIGURE 5.16
Software algorithm for the seven-segment display circuit shown in Figure 5.15.

this pointer is increased and the information stored in Tables 1 and 2 is sent to ports A and B.

With this approach, Table 1 can be considered as the RAM image of the display. Each piece of information that has to be displayed must be first written in this table. Before doing this it is necessary to convert the original information that can be in BCD (binary-coded decimal), ASCII, or a nonstandard code into the seven-digit code used by the displays.

5.3.4 Alphanumeric Liquid-Crystal Displays

Liquid-crystal displays (LCDs) are widely used to represent alphanumeric information. Most of them consist of the LCD screen itself and a microcontroller used to control the information that the screen displays through an easy-to-use hardware and software interface. These screens normally have one or two lines, each one able to display a certain number of characters. A very commonly used device is HD44780 from Renesas (formerly Hitachi). This microcontroller can directly handle an LCD screen of one or two lines, each containing eight characters. To handle larger screens, it is necessary to use auxiliary circuits such as the HD44100 driver that allows handling of an additional eight characters for each line. Figure 5.17 shows the block diagram for the LCD LM016L that uses a HD44780 controller and a HD44100 driver. This module controls an LCD of two lines with 16 characters per line.

The HD44780 microcontroller has an internal data memory (DDRAM) that can store the ASCII codes for up to 80 alphanumeric characters. These characters can be displayed in one or two lines. Each line operates like a circular memory, that is, once one line is full the next character is written in the initial memory position. If the circular RAM is organized in a single

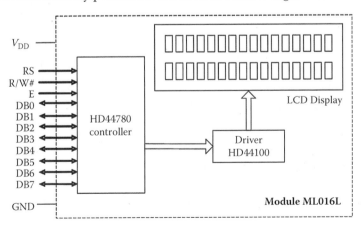

FIGURE 5.17
Components for LCD module ML016L with an 8-bit interface.

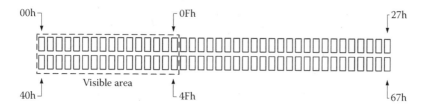

FIGURE 5.18
The data RAM memory (DDRAM) for the HD44780 controller can be organized as one or two lines. When organized as two lines (as shown in this figure), each line acts like a 40-byte circular memory with the addresses shown. The noncontinuous line shows the part of DDRAM visible in the LM016L module (two lines with 16 characters per line). The visible part can be moved along the DDRAM using the appropriate orders shown in Table 5.6.

line, it operates as an 80-byte circular memory with addresses that are sequential from 00h to 4Fh. If it is organized in two lines, it has two independent circular memories, each one of them able to store 40 characters. In this case, the memory addresses are not sequential; the addresses for the first line are from 00h to 27h and the addresses for the second line are from 40h to 67h, as shown in Figure 5.18.

The HD44780 also has a character generator in ROM and another one in RAM. The ROM generator stores the matrix of dots needed to generate each character. The RAM generator (CGRAM) allows the user to define nonstandard characters that are not stored in the ROM character generator. This microcontroller also has an instruction set to manipulate the display.

The main specifications for LCDs that use the HD44780 controller are:

- Ability to connect to 4-bit or 8-bit parallel ports
- Possibility to store up to 80 characters in an 80-byte internal RAM (DRAM)
- Character generator in internal ROM with:
  - 160 characters with 5×7 dots
  - 32 characters with 5×10 dots
- User-defined character generator in internal RAM with:
  - 8 characters with 5×7 dots
  - 4 characters with 5×10 dots
- CMOS technology, yielding very low power consumption and therefore able to be battery powered
- Internal power-on RESET circuit
- Large instruction set: clear display, cursor and character blinking, movement of cursor and display, on/off control for cursor and display, etc.

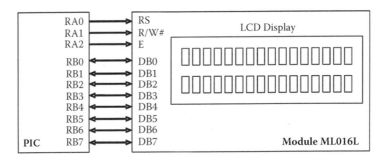

FIGURE 5.19
LCD module connected to a PIC microcontroller using ports A and B.

These LCD modules have a digital interface for the transfer of data and control signals between the module and the microcontroller or microprocessor. This interface consists of three control lines and four or eight data lines, depending on whether the interface is 4 or 8 bits.

The connection between a medium-end PIC microcontroller and a LCD module can be done by using ports A and B. Figure 5.19 shows the connection between these modules using an 8-bit interface. The lines and signals involved in this connection are described in the following list.

- Register select (RS). This signal tells the LCD module whether the signal being sent through DB0-DB7 is a control signal (RS = 0) or a data signal (RS = 1).
- Read/write (R/W). This signal indicates reading (RW = 1) or writing (RW = 0).
- Enable (E). E = 1 enables the device. The LCD module captures data or control signals with the falling edge of E.
- Data bus (DB0-DB7). Bidirectional lines that transmit data and control signals.

Figure 5.20 illustrates the process of sending a control or a data signal to the LCD module. Table 5.6 shows the order set accepted by LCD modules that use the HD44780 controller.

The HD44780 controller has an internal register called the address counter (AC). This register contains the address of DDRAM or CGRAM that will be used to write or read data. Once one type of memory has been selected, data will be read or written from that memory until the selection is changed.

The control instructions "Clear display" and "Move to initial position" select the DDRAM and set the AC to 0. The instruction "Select DDRAM" allows for setting the AC to any value. When data is written (or read) the AC register is incremented or decremented depending on the input mode selected.

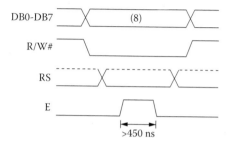

FIGURE 5.20
Signals used to write data or orders in the LCD module. Order writing (RS = 0) or data writing (RS = 1) becomes effective in sync with the falling edge of the signal E.

In general, the execution of orders for the display of data takes a relatively long time, on the order of several microseconds. During this time the controller cannot receive a new order or data. For this reason, before sending new data or orders it is necessary to find out if the controller can receive them. The busy flag (BF) bit is used for this reason; BF = 1 when the controller is busy and BF = 0 when the controller is ready to accept new data or new orders. This bit can be read at any time by reading (R/W# = 1) an order (RS = 0). Bit 7 in the returned word (DB0 to DB7) from the controller is the BF indicator. Example 5.6 illustrates this process.

Example 5.6

This example shows the assembler code for some basic routines to operate an LCD that incorporates the HD44780 controller. In this example the LCD module is connected to the PIC microcontroller through ports A and B as shown in figure 5.19

This example has four routines: INIT_LCD for the display initialization, WR_CMD and WR_DATA to write orders and data in the display, and LCD_BUSY to use the BF bit to check if the controller is busy before sending a new order or data.

```
        list       p = 16f873
        #include <p16f873.inc>
; Hardware description:
P_DATA   equ    PORTB        ; Port for display data lines.
P_TRIS   equ    TRISB
P_CTRL   equ    PORTA        ; Port for display control lines.
RS       equ    0            ; Control bit for signal RS.
RW       equ    1            ; Control bit for signal RW.
E        equ    2            ; Control bit for signal E.
; Other Declarations:
TEMP     equ    0x020        ; Temporary register used by subroutines.
; INIT_LCD: Display initialization subroutine.
INIT_LCD:
                             ; BF flag not ready yet.
        clrf   P_CTRL        ; Control line to 0.
```

TABLE 5.6

Orders Accepted by HD44780

Instruction	RS	R/W	DB7	DB6	DB5	DB4	DB3	DB2	DB1	DB0	Description
Clear display	0	0	0	0	0	0	0	0	0	1	Stores 20h (ASCII code for space character) in DDRAM, selects it and sets AC to 0.
Move to initial position	0	0	0	0	0	0	0	0	1	X	Selects DDRAM and sets AC to 0. Moves display to this initial position. Content in DDRAM is not modified.
Select input mode	0	0	0	0	0	0	0	1	I/D	S	Selects movement direction for cursor (I/D). Also selects whether the display moves or not (S). These operations can be done during data reading or writing.
Control display	0	0	0	0	0	0	1	D	C	B	Turns display (D) and cursor (C) on or off. Activates cursor blinking (B).
Move display or cursor	0	0	0	0	0	1	S/C	R/L	X	X	Moves display or cursor (S/C) in one direction (R/L).
Select function	0	0	0	0	1	DL	N	F	X	X	Selects size (4 or 8) for data bus (DL), the number of display lines (N), and character format (F).
Select CGRAM	0	0	0	1	CGRAM address that will be stored in AC						Stores a CGRAM address in AC. After this order, read or written data come or go to CGRAM.
Select DDRAM	0	0	1	DDRAM address that will be stored in AC							Stores a DDRAM address in AC. After this order, read or written data come or go to DDRAM.
Read BF and AC	0	1	BF	AC content							Reads the status of busy flag BF and the content of AC.
Write data	1	0	Data to write								Writes data in the address pointed by AC. Data is written in DDRAM or CGRAM depending on the last selection.
Read data	1	1	Read data								Reads data from the address pointed by AC. Data is read from DDRAM or CGRAM depending on the last selection.

Note: I/D—1: increment, 0: decrement; S—1: automatic displacement of display; S/C—1: display displacement, 0: cursor movement; R/L—1: right displacement, 0: left displacement; DL—1: 8-bit interface, 0: 4-bit interface; N—1: 2-lines display, 0: 1-line display; F—1: 5 × 10 dots characters, 0: 5 × 7 dots characters; BF—Busy flag, 1: display busy, 0: display can accept orders or data; DDRAM, data RAM memory; CGRAM, character generator RAM memory; AC, address counter.

```
        call DELAY15     ; Wait 15 ms.
         ; BF flag ready from here.
        movlw 38h        ; 2-lines display. 5 × 7 characters.
        call WR_CMD
        movlw 08h        ; Turn off display and cursor.
        call WR_CMD
        movlw 01h        ; Clear display and set AC = 0.
        call WR_CMD
        movlw 0Ch        ; Turn on display with cursor off.
        call WR_CMD
        movlw 06h        ; Select input mode: Cursor moves towards
                         ; right.
        call WR_CMD
        return
; WR_CMD: Subroutine to write an order in the display.
;       Input: Order must be stored in W.
WR_CMD:
        movwf TEMP       ; Store order in TEMP.
        call LCD_BUSY    ; Wait for display ready.
        bcf   P_CTRL, RW ; Prepare writing (RW = 0)
        bcf   P_CTRL, RS ; an order (RS = 0).
        bsf   P_CTRL, E  ; Enable display (E = 1).
        movf  TEMP, W    ; Store order W.
        movwf P_DATA     ; Send order to display.
        bcf   P_CTRL, E  ; Disable display (E = 0).
        return
;WR_DATA:Subroutine to write data in the display.
;       Input: Order must be stored in W.
WR_DATA:
        movwf TEMP       ; Store order in TEMP.
        call LCD_BUSY    ; Wait for display ready.
        bcf   P_CTRL, RW ; Prepare writing (RW = 0)
        bsf   P_CTRL, RS ; data (RS = 1).
        bsf   P_CTRL, E  ; Enable display (E = 1).
        movf  TEMP, W    ; Store data in W.
        movwf P_DATA     ; Send data to display.
        bcf   P_CTRL, E  ; Disable display (E = 0).
        return
;LCD_BUSY: Subroutine for waiting if display is busy.
;       This subroutine checks bit BF. Wait while BF = 1
;       and returns when BF = 0.
LCD_BUSY:
        bsf   STATUS,RP0 ; Select register bank 1.
        movlw 0FFh       ; Store data port in input
        movwf P_TRIS     ; by writing FFh in the appropriate TRIS
                         ; register
        bcf   STATUS, RP0; Select bank 0.
        bsf   P_CTRL, RW ; Prepare reading (RW = 1).
        bcf   P_CTRL, RS ; of an order (RS = 0).
BUSY10:
        bsf   P_CTRL, E  ; Enable display (E = 1).
        nop
        movf  P_DATA, W  ; Read display. BF is bit 7.
        bcf   P_CTRL, E  ; Disable display (E = 0).
        andlw 80h        ; Retrieve BF bit.
        btfss STATUS, Z  ; Is BF 0?
        goto  BUSY10     ; No - display busy. Check again.
        bcf   P_CTRL, RW ; Yes - display not busy. End of reading
                         ; (RW = 1).
        bsf   STATUS, RP0; Select bank 1.
        movlw 0          ; Store data port in output.
```

```
        movwf P_TRIS      ; by writing 00h in the
                          ; appropriate TRIS register.
        bcf   STATUS, RP0; Select bank 0.
        return            ; Return.
; DELAY15: Subroutine for 15 ms delay.
DELAY15:
    ;
    ; Write here code for this subroutine.
    ;
    return
    end
```

6

Timers

Many microcontroller applications, such as generating signals, measuring the duration of a signal, and keeping date and time, use time as their variable. For this reason, microcontrollers need to have internal resources to accurately measure time. Each PIC microcontroller has at least one basic timer module called Timer0, although the majority of medium-end PICs can have two additional timers: Timer1 and Timer2. In addition, some PICs have capture/compare/PWM (CCP) modules that increase the possibilities of the basic timers. This chapter describes the structure, function, and programming of each one of these modules, illustrated with examples of how they work and how they are programmed.

6.1 Timers in PIC Microcontrollers

Each one of the timers in a medium-end PIC microcontroller is based on an 8-bit or 16-bit incrementing synchronous counter. These counters can be programmed to count internal or external pulses. The count number stored by each counter can be read or modified by accessing the special function registers associated with that timer. Some of the bits in these registers are used to notify of counter overflow, being able to generate an interrupt request to the microcontroller.

Timers can also have an auxiliary asynchronous counter. This auxiliary counter can be configured as a prescaler (when placed in the signal path before the main counter) or as a postscaler (after the main counter). Timer0 and Timer1 only have prescalers, whereas Timer2 has a prescaler and a postscaler. Figure 6.1 shows the general schematic for the modules Timer0, Timer1, and Timer2. The main specifications of these timers are shown in table 6.1 and are described next in further detail.

6.1.1 Timer0 Module

Timer0 consists of a prescaler and an 8-bit incrementing synchronous counter that can be read or written using the special function register TMR0 as shown in Figure 6.2. The prescaler is an asynchronous coun-

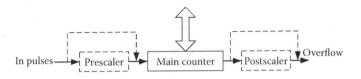

FIGURE 6.1

General block diagram for medium-end PIC timers. All of them have an 8-bit or 16-bit incrementing main counter and a prescaler. Timer2 also has a postscaler. The pulses to be counted can be internal or external. Overflow is reported in a bit from the special function registers. This bit can generate an interrupt request to the PIC.

TABLE 6.1

Main Specifications for Timers in Medium-End PIC Microcontrollers

Timer	Size	Prescaler Division	Postscaler Division	SFR for Count Number	Overflow
Timer0	8 bits	2, 4, ..., 256	NO	TMR0	bit T0IF in OPTION
Timer1	16 bits	1, 2, 4, 8	NO	TMR1H, TMR1L	bit TMR1IF in PIR1
Timer2	8 bits	1, 4, 8	1, 2, ..., 16	TMR2	bit TMR2IF in PIR2

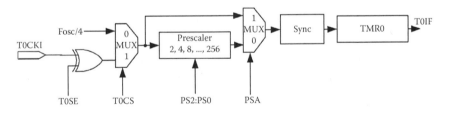

FIGURE 6.2

Block diagram for Timer0. Timer0 can count machine cycles (timer mode, T0CS = 0) or external pulses (counter mode, T0CS = 1). TMR0 is an 8-bit incrementing counter. When TMR0 overflows, it activates the flag T0IF in the INTCON register. Before reaching TMR0, the pulses are synchronized with the microcontroller's clock. They might be affected by a programmable prescaler.

ter with a programmable division factor. The count of this prescaler is invisible to the programmer.

Timer0 can be configured to count machine cycles or to count external pulses. When counting machine cycles it is said to operate as a timer. When counting external pulses, it is said that it operates as a counter. The external pulses are connected to the T0CKI pin. When the pulses reach the synchronization block, they are sampled twice during each machine cycle. This results in a new signal whose edges are in phase with the microcontroller's clock. This synchronized signal is used to drive the counter TMR0. In order to not lose pulses during the synchronization, it is necessary for the pulses that enter the block to remain at 1 or 0 at least half of the duration of the machine cycle. When Timer0 works in counter mode, the synchronization block determines the minimum value of the period for the pulses that enter through the T0CKI pin. If T_{OSC} is the period for the main oscillator in the microcontroller and P is the prescaler factor, the period Ti of the pulses entering through the T0CKI must meet the following condition:

$$Ti > \frac{4 \times T_{OSC}}{P} \, , \qquad (6.1)$$

where $P = 1$ when the prescaler is not used and $P = 2, 4, \ldots, 256$ when it is.

In low-power mode (sleep), the main oscillator stops working, resulting in Timer0 not working when the microcontroller is sleeping.

There are three special function registers associated with Timer0: TMR0, OPTION, and INTCON. Figure 6.3 shows the names of the bits in the OPTION and INTCON registers. The TMR0 register stores the value for the counter in Timer0. This value can be read or written at any time from the program executed by the microcontroller. When a value is written in TMR0, the count for the prescaler —if the prescaler is assigned to the timer—is set to 0. Also, writing in the TMR0 register inhibits the counting in Timer0 during two machine cycles.

When Timer0 overflows, the flag T0IF (bit INTCON<2>) is set to 1. If Timer0 is serviced using programmed I/O, this bit must be checked to learn if Timer0 did overflow. Once the overflow is verified, this flag can be reset by software. The interrupt to Timer0 can be enabled by setting bit

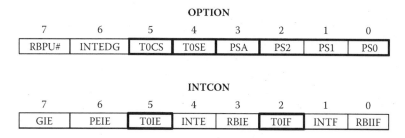

OPTION

7	6	5	4	3	2	1	0
RBPU#	INTEDG	T0CS	T0SE	PSA	PS2	PS1	PS0

INTCON

7	6	5	4	3	2	1	0
GIE	PEIE	T0IE	INTE	RBIE	T0IF	INTF	RBIIF

FIGURE 6.3

Special function registers OPTION and INTCON. OPTION stores the configuration bits for Timer0: T0CS configures Timer0 as a timer or counter; T0SE configures the external signal edge that will increment Timer0 when working as a counter; PSA assigns the prescaler to Timer0 or the watchdog; and PS2, PS1, and PS0 program the division factor for the prescaler. INTCON stores the Timer0 overflow flag, T0IF. Bit T0IE enables the interrupt request to the microcontroller as a result of Timer0 overflow.

T0IE (INTCON<5>) to 1. If this interrupt is enabled, Timer0 generates an interrupt request when it overflows.

The control bits for Timer0 are in the OPTION register. Bit T0CS selects the source for the clock pulses. When selecting an external clock source at the T0CKI pin, bit T0SE selects if the counter increments with the raising edges (T0SE = 0) or with the falling edges (T0SE = 1) of the pulses in T0CKI.

As shown in Figure 6.4, Timer0 and the watchdog timer (WDT) share the prescaler. This prescaler is an 8-bit asynchronous counter that can be assigned to Timer0 or to the WDT. When assigned to one module, the other module cannot use it. The prescaler is assigned to Timer0 by setting the bit PSA (OPTION<3>) to 1. If the bit PSA is set to 0, the prescaler is assigned to the WDT. Bits PS2, PS1, and PS0 in the OPTION register select the division factor for the prescaler.

The division factor (P) for the prescaler assigned to Timer0 can have the following values:

$$P = 2, 4, \ldots, 2^{n+1}, \ldots, 256, \tag{6.2}$$

with $n = 0, 1, \ldots, 7$ being the value stored in bits PS2:PS0.

When the prescaler is assigned to the WDT, the division factor is $P = 1$, $2, 4, \ldots, 2^n, 2^{n+1}, \ldots, 128$, as shown in table 2.1.

The overflow time for Timer0 can be calculated as follows. Let N be the number of pulses that need to reach Timer0 to overflow it, P the prescaler division factor, and Ti the period of the pulses at the input of the prescaler. If Timer0 works as a timer, then Ti is the duration of a machine cycle; if it

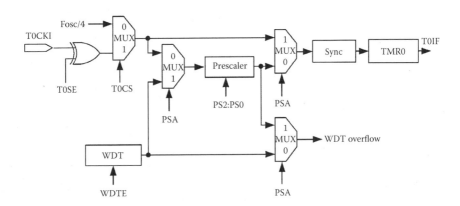

FIGURE 6.4
The prescaler can be assigned to Timer0 or to the watchdog timer (WDT). PSA = 0 assigns it to Timer0 and PSA = 1 assigns it to WDT. When Timer0 overflows (moves from FFh to 00), the bit T0IF in the register INTCON becomes active. T0SE, T0CS, PSA, and PS2:PS0 are bits for the OPTION register. WDTE is one of the PIC configuration bits that enables the WDT.

operates as a counter, *Ti* is the period of the external pulses. The overflow time (*To*) can be found as

$$To = P \times N \times Ti. \tag{6.3}$$

Note that the value stored in the TMR0 register is not *N* but its 2-complement using 8 bits, which is the value that *N* needs to reach 256. Therefore, the value that must be loaded in TMR0 is

$$N_{TMR0} = 256 - N. \tag{6.4}$$

If Timer0 works as a timer (counting machine cycles) it is necessary to keep in mind that the counting is inhibited during two machine cycles after writing data in the TMR0 register. Therefore, the value that needs to be loaded in the TMR0 is

$$N_{TMR0} = 256 - N + 2. \tag{6.5}$$

Finally, it is important to realize that if the value N_{TMR0} is loaded in the TMR0 register each time there is an overflow, then Timer0 works as a *N*-module counter or timer.

Example 6.1

Use of Timer0 to program delays. To carry out this task, Timer0 operates as a timer because its clock pulses come from the main oscillator. If f_{osc} is the frequency of the main oscillator, the frequency of the pulses reaching the timer is $f_{osc}/4$. The following program illustrates how to initialize the operation of Timer0 (subroutine InitTimer0) and how to assign the prescaler to Timer0 with a division factor equal to 8. The subroutine Del1ms creates a delay of approximately 1 ms. The subroutine DelNms creates a delay of *N* ms with *N* being less than or equal to 255.

```
; Using Timer0 and prescaler to program delays.
;
; Hardware:
; Oscillator frequency for PIC: 4 MHz, therefore, a machine
; cycle (MC) lasts Tmc = 1 µs.
;
; Values to store in prescaler and TMR0 for a 1 ms delay:
; 1 ms = 1000 µs, but 1000 = 8 x 125, therefore store in
; prescaler
; P = 8 and in TMR0 the 2-complement of 125 plus 2.
; That is, TMR0 = 256 - 125 + 2 = 133.
      List         p = 16F873
      include  "P16F873.INC"

AUX    equ          0x20              ; Auxiliary variable
; InitTimer0: Subroutine to program Timer0 as timer with a
   prescaler of 8.
```

```
;
InitTimer0:
      bcf         INTCON, T0IE     ; Disable Timer0 interrupt.
      bsf         STATUS, RP0      ; Select data memory bank 1
      movlw       0xC2             ; and configure Timer0 as a
                                   ; timer
      movwf       OPTION_REG       ; with prescaler factor of 8.
      bcf         STATUS, RP0      ; Select data memory bank 0.
      clrf        TMR0             ; Store 0 in TMR0.
      bcf         INTCON, T0IF     ; Set overflow flag to 0.
;     bsf         INTCON, T0IE     ; It is possible to enable
                                   ; here Timer0 interrupt
                                   ; if necessary.

      return
; Del1ms: Subroutine for a 1 ms delay
;      Inputs: none.
;      Outputs: none.
Del1ms:
      movlw       .133             ; 125 in 2-complement plus 2,
      movwf       TMR0             ; stored in TMR0.
Del1ms_01:
      btfss       INTCON, T0IF     ; T0IF = 1?
      goto        Del1ms_01        ; No - wait.
      bcf         INTCON, T0IF     ; Yes - set T0IF = 0 and
      return                       ; return because 1 ms has
                                   ; elapsed.
; DelNms: Subroutine to create a delay of N milliseconds (N
;     < = 255).
;         This subroutine calls N times subroutine Del1ms.
;         Inputs: N in W.
;         Outputs: none.
DelNms:
      movwf AUX                    ; Store the delay (ms) in AUX.
DelNms_01:                         ; Call Del1ms N times.
      call        Del1ms           ; Wait 1 ms.
      decfsz      AUX, f           ; Decrement AUX. AUX = 0?.
      goto        DemNms_01        ; No - continue waiting.
      return                       ; Yes - return because N ms
                                   ; have already elapsed.

      end
```

6.1.2 Timer1 Module

Timer1, whose structure is shown in Figure 6.5, is the second timer module available in most medium-end PIC microcontrollers. Timer1 consists of a 16-bit incrementing counter with a prescaler with a division factor of 1, 2, 4, or 8. Timer1 can work as a timer (counting machine cycles) or as a counter (counting external pulses). When working as a counter, Timer1 can be programmed to work in synchronous or asynchronous mode. This selection is done by the T1SYNC# bit in the T1CON register. When T1SYNC# = 0, Timer1 operates as a synchronous counter because the input pulses to TMR1 travel through the synchronization block. This block samples the input signal and synchronizes it with the microcontroller's internal clock. The resulting signal has edges that are in phase with the main clock of the PIC. This synchronized signal drives the 16-bit counter TMR1 that is made by the TMR1L and TMR1H registers.

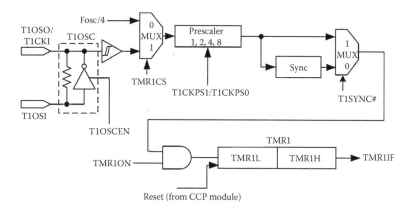

FIGURE 6.5
Timer1 block diagram.

To not lose pulses during the process of synchronization, it is necessary that all the pulses entering the synchronization block remain at 1 or 0 at least half of the duration of a machine cycle. This determines the minimum period of the pulses that enter in Timer1 in this operating mode. The working modes synchronized and nonsynchronized apply only to Timer1 working as a counter, because when it works as a timer it always works in synchronized mode. When Timer1 is configured to work as a timer by setting TMR1CS = 0, the T1SYNC# bit is ignored.

If T_{OSC} is the period of the main oscillator in the microcontroller and P is the division factor for the prescaler ($P = 1, 2, 4, 8$), the period Ti for the pulses in T1CKI must be

$$Ti > \frac{4 \times T_{OSC}}{P}. \tag{6.6}$$

With Timer1 programmed as an asynchronous counter (T1SYNC# = 1) this counter continues working even when the microcontroller is in low-power consumption mode. This makes Timer1 an excellent candidate for a real-time clock (RTC). The count value can be written or read using the special function registers TMR1H and TMR1L. When these registers are written, the count for the prescaler is reset to 0. The T1CON register contains the control bits for Timer1. The source of its clock pulses can be internal or external, selected using the TMR1CS bit. The external clock can consist of the pulses entering through the T1CKI/T1OSC pin or from an external crystal set across pins T1OSO and T1OSI. The T1OSCEN bit enables the oscillator to use an external crystal. Bit TMR1ON enables the counting process.

The value n ($n = 0, 1, 2, 3$) for the bits T1CKPS1 and T1CKPS0 in the T1CON register sets the division factor P ($P = 1, 2, 4, 8 = 2^n$) for the prescaler

T1CON

7	6	5	4	3	2	1	0
–	–	T1CKPS1	T1CKPS0	T1OSCEN	T1SYNC#	TMR1CS	TMR1ON

FIGURE 6.6
T1CON: Timer1 control register.

in Timer1, as shown in Figure 6.6. For example, if a division factor of 8 is desired, these bits will be set to 3. When Timer1 overflows, the TMR1F bit is set to 1. TMR1F is a bit in the PIR1 register, and after overflowing it must be reset to 0 by software. If the interrupt for Timer1 is enabled (this is done by setting bit TMR1E to 1), the overflow generates an interrupt.

It is possible to calculate the time needed for Timer1 to overflow by following the same reasoning as for Timer0. With N being the number of pulses needed before the 16-bit counter (TMR1) will overflow, P the division factor for the prescaler, and Ti the period for the input pulses, the overflow time for Timer1 is

$$To = P \times N \times Ti. \tag{6.7}$$

The 2-complement for N using 16 bits is loaded in the registers TMR1L and TMR1H:

$$N_{TRM1} = 2^{16} - N = 65536 - N. \tag{6.8}$$

When Timer 1 runs freely, then $N = 2^{16}$. In this case,

$$To = P \times 2^{16} \times Ti. \tag{6.9}$$

Example 6.2

Timer1 programming. This example illustrates how to program Timer1 as an asynchronous counter for external pulses; how to write a 16-bit binary number in registers TMR1L and TMR1H; and how to safely read registers TMR1L and TMR1H.

```
        List      p = 16F873
        include   "P16F873.INC"
;
AUX_H equ      0x20              ; Auxiliary variable.
AUX_L equ      0x21              ; Auxiliary variable.
;
; InitTimer1: Routine to program Timer1 as an asynchronous
  counter for external pulses with a prescaler value of 8.
InitTimer1:
        clrf  T1CON              ; Timer1 as timer. Prescaler = 1.
```

```
        bsf     STATUS, RP0     ; Select bank 1.
        bcf     PIE1, TMR1IE    ; Disable interrupt for Timer1.
        bcf     STATUS, RP0     ; Select bank 0.
        clrf    TMR1H           ; Set TMR1H to 0.
        clrf    TMR1L           ; Set TMR1L to 0.
        bcf     PIR1, TMR1IF    ; Overflow flag to 0.
        movlw   0x36            ; Configure Timer1 as asynchronous
                                ; counter with P = 8.
        movwf   T1CON           ; T1OSC disabled, Timer1 stopped.
        bsf     T1CON, TMR1ON   ; Start Timer1.
        return
;
; WR_TMR1a: Routine to write a 16 bit binary number in Timer1.
;           Temporarily stopping Timer1 count.
;           Inputs: AUX_H and AUX_L contain the high and low
;           bytes for the number to write in Timer1.
;           Outputs: none
WR_TMR1a:
        bcf     T1CON, TMR1ON   ; Stop TMR1.
        movf    AUX_H, W        ; Load high byte in
        movwf   TMR1H           ; TMR1H register.
        movf    AUX_L, W        ; Load low byte in
        movlw   TMR1L           ; TMR1L register.
        bsf     T1CON, TMR1ON   ; Reinitiate counting TMR1.
        return
;
; WR_TMR1b: Routine to write a 16 bit binary number in Timer1.
;       without stopping Timer1 count.
;       Inputs: AUX_H and AUX_L contain the high and low bytes
;       for the number to write in Timer1.
;       Outputs: none
;
WR_TMR1b:
        clrf    TMR1L           ; Ensure TMRL1 will not overflow
                                ; while
        movf    AUX_H, W        ; loading high byte on
        movwf   TMR1H           ; register TMR1H.
        movf    AUX_L, W        ; Load low byte in
        movlw   TMR1L           ; register TMR1L.
        return
;
; RD_TMR1: Routine to read value of Timer1 while it is counting.
;           Inputs: None.
;           Outputs: In AUX_H return the value of TMR1H and
;           AUX_L returns the value of TMR1L.
;
RD_TMR1:
        movf    TMR1H, W        ; Read TMR1H and store high byte
        movwf   AUX_H           ; in AUX_H.
        movf    TMR1L, W        ; Read low byte in TMR1L and
                                ; store it
        movwf   AUX_L           ; in AUX_L.
        movf    TMR1H, W        ; Read again TMR1H to check if
                                ; changed.
        xorwf   AUX_H, W        ; Compare readings.
        btfsc   STATUS, Z       ; Same?
        goto    RD_END          ; Yes - Reading validated. Finish
                                ; routine.
        movf    TMR1H, W        ; No - Non-valid reading. Read
                                ; again TMR1H
        movwf   AUX_H           ; and TMR1L. These results are
                                ; valid reading because
```

```
              movf  TMR1L, W       ; there is no time for TMR1L to
                                   ; overflow and change
              movwf AUX_L          ; the value of TMR1H.
    RD_END:
              return               ; Return with the correct readings.
      ;
      end
```

6.1.3 Timer2 Module

Timer2 is a third module available in some medium-end PIC microcontrollers. It consists of an 8-bit incrementing counter, a prescaler, a postscaler, and a register to store the count. Timer2 can only work as a timer counting machine cycles. Figure 6.7 shows its block diagram.

The prescaler can be programmed with division factors of 1, 4, or 8. The values for the postscaler are 1, 2, 3, ..., 16. When using both scalers with their maximum values, the overflow time for Timer2 is the overflow time for a 16-bit counter. Timer2 is the timer that generates the time base used by the pulse width modulator when the CCP module operates in PWM mode. Timer2 can also be used to generate the clock for the SSP module. Timer2 does not increment while the microcontroller is in low-power mode. It restarts counting when the microcontroller wakes up.

Figure 6.8 shows the bits that make up the control register for Timer2, T2CON. Bits T2CKPS1 and T2CKPS0 select the division factor for the prescaler, while bits TOUTPS3:TOUTPS0 select the division factor for the postscaler. The values for these bits are shown in Table 6.2. Timer2 is enabled by setting bit TMR2ON in register T2CON to 1 and is disabled by setting this bit to 0.

The count value can be read or written in register TMR2. The changing value in register TMR2 is continuously compared against the value stored in PR2. When these two values are equal, the register TMR2 is set to 0 in the next clock cycle and also sends a pulse to the postscaler. With P2 being the division factor in the postscaler, Timer2 overflows after P2

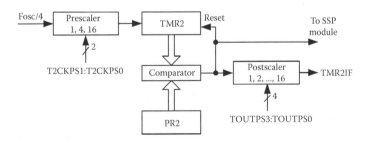

FIGURE 6.7
Timer2 block diagram.

T2CON

7	6	5	4	3	2	1	0
–	TOUTPS3	TOUTPS2	TOUTPS1	TOUTPS0	TMR2ON	T2CKPS1	T2CKPS0

FIGURE 6.8

T2CON: Timer2 control register. Bits T2CKPS1 and T2CKPS0 select the division factor for the prescaler, and bits TOUTPS3:TOUTPS0 select the division factor for the postscaler. Bit TMR2ON enables Timer2 counting.

TABLE 6.2

Bits in T2CON Register and Division Factors for Prescaler and Postscaler

T2CKPS1:T2CKPS0	Prescaler Division Factor	TOUTPS3:TOUTPS0	Postscaler Division Factor
00	1	0000	1
01	4	0001	2
1x	16	0010	3
	1111	16	

Note: The values T2CKPS1 and T2CKPS0 set the prescaler division factor, and the value of TOUTPS3:TOUTPS0 selects the postscaler division factor.

pulses. When this happens, bit TMR2IF in register PIR1 is set to 1. This bit must be reset to 0 by software. If bit TMR2IE in register PIE1 was set to 1, this process generates an interrupt request.

The overflow time for Timer2 can be calculated as follows: With N being the number stored in register PR2, P_1 the division factor for the prescaler, P_2 the division factor for the postscaler, and Ti the period of the pulses entering the module, the overflow time To for Timer2 is:

$$To = P_1 \times P_2 \times (N + 1) \times Ti, \tag{6.10}$$

in which $P_1 = 1, 4, 16$ and $P_2 = 1, 2, 3, \ldots, 16$. Considering that Timer2 acts like a timer,

$$Ti = 4 \times T_{OSC}. \tag{6.11}$$

Example 6.3

Program Timer2 to overflow every 1 ms using a main oscillator for the microcontroller with a frequency of 4 MHz.

With a 4 MHz frequency, the period of the pulses entering Timer2 is Ti = 1 μs. To achieve an overflow time Td = 1 ms, Timer2 has to count up to 1000. This can be seen using Equation 6.10:

$$\frac{Td}{Ti} = P_1 \times P_2 \times (N+1) = 1000.$$

This value can be achieved with P_1 = 4, P_2 = 10, and N = 24.

The following segment of code illustrates how to program Timer2 with these parameters.

```
      List       p = 16F873
      include  "P16F873.INC"
;
; InitTimer2: Subroutine that programs Timer2 to divide by 1000.
InitTimer2:
      clrf     T2CON            ; Stop Timer2.
      clrf     TMR2             ; Set TMR2 to 0.
      bsf      STATUS, RP0      ; Select bank 1.
      bcf      PIE1, TMR2IE     ; Enable interrupt for Timer2.
      movlw    .24              ; Count number for TMR2
      movwf    PR2              ; in PR2.
      bcf      STATUS, RP0      ; Select bank 0.
      bcf      PIR1, TMR2IF     ; Overflow flag is reset to 0.
      movlw    0x4A             ; Postscaler = 10,
                                ; Prescaler = 16.
      movwf    T2CON            ; Timer2 stopped.
      bsf      T2CON, TMR2ON    ; Timer2 starts counting.
      return
;
;
; Wait_Timer2: Subroutine to wait for Timer2 overflow
; that happens each 1 ms.
;
Wait_Timer2:
      btfss    PIR1, TMR2IF     ; Timer2 overlow?
      goto     Wait_Timer2      ; No - wait.
      bcf      PIR1, TMR2IF     ; Yes - Set flag TMR2IF to 0.
      return                    ; Return.
      end
```

6.2 The CCP Module

The capture/compare/PWM (CCP) modules are circuits that when used together with Timer1 and Timer2 allow for other forms of timing signals. A single microcontroller can have up to two CCP modules called CCP1 and CCP2. A CCP module consists of two 8-bit registers called CCPRxH and CCPRxL, with x being 1 or 2 depending on the CCP module to which they refer. These registers can store the high and low bytes of a 16-bit number. Each CCP module also uses the CCPxCON register for control, and the bit CCPxIF in the PIR register to indicate the presence of an event.

If the interrupt for the module is enabled (the interrupt can be enabled with the bit CCPxIE in the PIE register), it produces an interrupt request when CCPxIF is set to 1.

Each CCP module can operate in the following modes:

- Capture mode. The CCP module captures the value of Timer1 when an external event occurs in pin CCPx.

- Compare mode. The register in the CCP module stores a 16-bit number that is compared with the value in Timer1. The result of the compare process may generate an event that may include a change in the CCPx pin.

- Pulse width modulation (PWM) mode. The CCP module and Timer2 make up a PWM modulator whose output is located in pin CCPx.

The CCPx pins (CCP1 or CCP2) are inputs when the module operates in capture mode and outputs when it operates in compare or PWM modes. There is a CCP pin for each CCP module inside the PIC microcontroller. These pins share functions with the port C pins. Capture and compare modes use Timer1 as a time base. When using these modes, Timer1 must be programmed as a timer or as a synchronized counter. In PWM mode, Timer2, which always works as a timer, determines the frequency of the PWM signal.

Because the CCP modules share functions with timers Timer1 and Timer2, for those PIC microcontrollers with two CCP modules, such as the PIC16F873, it is necessary to keep in mind that both modules share the timers. Table 6.3 shows the possible interactions that must be considered when programming the CCP modules.

The special function registers CCPxCON are used to program the CCP modules. Figure 6.9 shows the bits of these registers. The mode of working for the CCP registers is programmed with bits CCPxM3:CCPxM0. Bits DCxB1 and DCxB0 are only used in PWM mode.

Table 6.4 shows the values that must be set in bits CCPxM3:CCPxM0 in the registers CCPxCON to program the different operation modes for the CCP modules. These operation modes are explained in further detail in the following sections.

6.2.1 Capture Mode

Figure 6.10 shows the block diagram for the CCP module working in capture mode. The value of Timer1 is stored in registers CCPRxH and CCPRxL when a specific event related to pin CCPx in the module occurs. The capture for Timer1 can be programmed to occur with the raising or falling edges of the input pulses in pin CCPx or with the rising edges every

TABLE 6.3

Interaction between CCP Modules

Mode for CCPx Module	Mode for CCPy Module	Interaction
Capture	Capture	Both modules use the same time base (Timer1).
Capture	Compare	The CCP module working as comparator must be configured to set Timer1 to 0 when the result of the comparison is positive.
Compare	Compare	The comparators must be configured to set Timer1 to 0 when the result of the comparison is positive.
PWM	PWM	Both PWM signals have the same period, given by the value in the PR2 register.
PWM	Capture	No interaction.
PWM	Compare	No interaction.

CCPxCON

7	6	5	4	3	2	1	0
−	−	DCxB1	DCxB0	CCPxM3	CCPxM2	CCPxM1	CCPxM0

FIGURE 6.9
CCPxCON registers used to control the CCP modules.

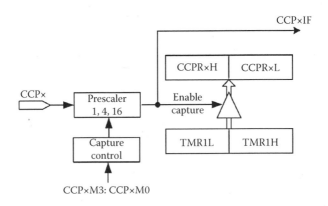

FIGURE 6.10
CCP module operating as a capture register for Timer1 for an event in pin CCPx.

TABLE 6.4

Values for CCPxM3:CCPxM0 Bits in the CCPxCON Register Associated with
Operation Modes for the CCP Modules

Bits CCPxM3:CCPxM0		Mode	
00	00	Disable CCP module.	
01	00	**Capture**	Each falling edge.
	01		Each rising edge.
	10		Each 4 rising edges.
	11		Each 16 rising edges.
10	00	**Comparator**	Pin CCPx is initiated low and is set to high when the result of the comparison is positive. Bit CCPxIF is set to 1.
	01		Pin CCPx is initiated high and is set to low when the result of the comparison is positive. Bit CCPxIF is set to 1.
	10		Bit CCPIF is set to 1 when the result of the comparison is positive. CCPx pin is not affected.
	11		Timer1 is set to 0 when the result of the comparison is positive. Bit CCPxIF is set to 1. CCPx pin is not affected.
11	xx	**PWM**	

4 or 16 pulses. This programming is done with bits CCPxM3:CCPxM0, as shown in Table 6.4. After a capture, bit CCPxIF is set to 1. This event can be used as a flag by the program. Also, if the interrupt to the CCP module is enabled (bit CCPxIE in register PIE is set to 1), an interrupt request is generated. The CCPx pin must be configured as an input pin by setting to 1 the appropriate bit in the TRIS register from the parallel port in which the CCPx is located, which is normally port C.

Example 6.4

The capture mode in the CCP module can be used to measure time. This example illustrates how to measure the period of a train of pulses using the CCP1 module in a PIC16F873. The frequency of the main oscillator in this example is 4 MHz.

Figure 6.11 shows the simplified circuit being used. This consists of the CCP1 module working in capture mode and Timer1. The main elements are the CCP1 pin, the signal period (T_x) of which we want to measure the CCPR1 registers (made of CCPR1H and CCPR1L); the TMR1 register (made of registers TMR1H and TMR1L); and the prescalers for the CCP module and Timer1 with division factors P_c and P_1, respectively.

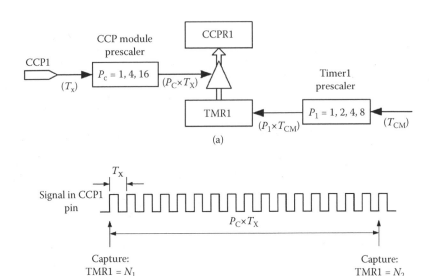

FIGURE 6.11
CCP module in capture mode. (a) Simplified block diagram of CCP module with Timer1.
(b) Time diagram showing two consecutive captures as a function of the period from the
train of pulses T_x in pin CCP1, period of a machine cycle T_{MC}, and the division factors for
the prescalers.

To ensure the capture mode works correctly, Timer1 must be programmed
as a timer to count machine cycles. Each machine cycle lasts T_{MC}. From the
schematic shown in figure 6.11 it can be seen how the capture of the changing
value of Timer1 occurs every $P_c \times T_x$ seconds. With N_1 and N_2 being the values
for TMR1 in two consecutive captures, Timer1 has changed in $N = N_2 - N_1$. In
time units this means that the seconds elapsed between the two consecutive
captures is $N \times P_1 \times T_{MC}$ seconds. Therefore,

$$P_c \times T_x = N \times P_1 \times T_{MC}$$

and

$$T_x = N \times \frac{T_{MC}}{P_c / P_1}.$$

The ratio P_c/P_1, which is the ratio of the division factors in the prescalers,
can be used to adjust the resolution with which T_x is measured, because in
a microcontroller the length of a machine cycle is already fixed. As $P_c = 1, 4,$
16 and $P_1 = 1, 2, 4, 8$, the best resolution is achieved with $P_c = 16$ and $P_1 = 1$.
These values allow for measuring the period T_x with a resolution equal to 1/16

of the length of a machine cycle. For this to be true, it is necessary, however, for the input signal to remain stable at least during the 16 periods that are used to measure it.

On the other hand, if $P_c = P_1$, then N represents the amount of machine cycles for the period T_x. If $T_{MC} = 1$ μs, then N is directly the length of the period T_x in microseconds. Because N is a 16-bit number, the longest period that can be measured using this method is 65536 μs. We leave it for the reader to calculate the maximum period that can be measured if $P_c/P_1 = 16$.

The following segment of code illustrates how to program Timer1 and the CCP1 module in capture mode to measure the period of pulses connected to pin CCP1 in a PIC16F873. The frequency of the main oscillator is 4 MHz and the division factors that have been chosen are $P_c = P_1 = 1$. With these values, the difference between two consecutive captures is equal to the period to be measured expressed in microseconds.

```
            List        p = 16F873
            include  "P16F873.INC"
N1H         equ      20h             ; High part of first capture.
N1L         equ      21h             ; Low part of first capture.
NH          equ      22h             ; High part of difference.
NL          equ      23h             ; Low part of difference.
; Init_capture: Subroutine to program module CCP1 in capture
; mode
; with raising edges for the pulses at the CCP1 pin. Timer1 has
; been
; programmed as timer with prescaler = 1.
Init_capture:
            clrf     T1CON           ; Timer1 as timer,
                                     ; prescaler = 1,stop.
            clrf     CCP1CON         ; Reset module CCP1.
            bsf      STATUS, RP0     ; Select bank 1.
            bsf      TRISC, 2        ; Set CCP1 pin as an input.
            bcf      PIE1, TMR1IE    ; Disable Timer1 interrupt.
            bcf      PIE1, CCP1IE    ; Disable CCP1 interrupt.
            bcf      STATUS, RP0     ; Select bank 0.
            clrf     PIR1            ; Set interrupt flags to 0.
            movlw    0x05            ; Select capture mode with each
                                     ; raising edge.
            movwf    CCP1CON         ;
            bsf      T1CON, TMR1ON   ; Start Timer1 counting.
            return
; Capture: This subroutine captures 2 values for Timer1 and
; calculates their difference.
; The difference, a 16 bit number is returned.
; in register NH (High part) and NL (low part).
Capture:
            bcf      PIR1, CCP1IF    ; Set capture indicator to 0.
            btfss    PIR1, CCP1IF    ; CCP1IF = 1?
            goto     Capture         ; No - wait.
            bcf      PIR1, CCP1I     ; Yes - Set capture indicator to
                                     ; 0 and
            movf     CCPR1L, W       ; store the captured value in N1H
                                     ; and N1L.
            movwf    N1L
            movf     CCPR1H, W
            movwf    N1H
Capture2:                           ; Capture next value:
            btfss    PIR1, CCP1IF    ; CCP1IF = 1?
```

```
        goto    Capture2     ; No - wait
        bcf     PIR1, CCP1IF ; Yes - Set capture indicator to
                             ; 0 and
                             ; subtract the captured values.
; This carries out the operation CCPR1 - N1 ==> N:
        movf    N1L, W
        subwf   CCPR1L, W
        movwf   NL
        btfss   STATUS, C
        goto    Subt1
        goto    Subt0
Subt1:
        decf    CCPR1H, f
Subt0:
        movf    N1H, W
        subwf   CCPR1H, W
        movwf   NH
        return
        end
```

6.2.2 Compare Mode

Figure 6.12 shows the block diagram for the CCP module working in compare mode. This mode compares the changing value of Timer1 with the value stored in registers CCPRxH and CCPRxL. When these values are equal, the result of the compare is positive and the module generates a specific event. The nature of this event can be programmed by using bits CCPxM3:CCPxM0 in the CCPxCON register as shown in table 6.4.

The events that can be programmed are to set pin CCPx to 0 or 1, and the reset for Timer1. When using the CCPx pin, it must be configured as an output by setting to 0 the appropriate bit in the TRIS register for the parallel port where the CCPx pin is located, usually port C. In all cases, when the result of the comparison is positive, bit CCPxIF in the PIR register is set to 1. This bit can be checked by the program. If the CCP module

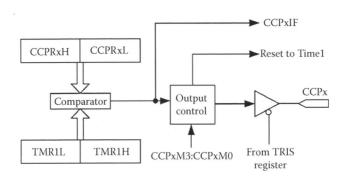

FIGURE 6.12
CCP module as comparator. When the value in the CCPRx register is equal to the value in the TMR1 pair, an event is generated in pin CCPx. The characteristics of the event are programmed using bits CCPxM3:CCPxM0 from the CCPxCON register.

interrupt is enabled (by setting bit CCPxIE in the PIE register to 1), it generates an interrupt request.

One of the events that can be generated as the result of the comparison being positive is to reset Timer1. This option increases the possibilities for Timer1 as it can work as a 16-bit comparator with a count module equal to the value stored in the CCPRxH and CCPRxL registers in the CCP module.

Example 6.5

Using Timer1 as a 16-bit timer with registers CCPR1H and CCPR1L storing the count module. For this configuration, the CCP1 module is programmed in compare mode, generating a reset for Timer1 each time that the comparison between registers CCPR1 (CCPR1H:CCPR1L) and TMR1 (TMR1H:TMR1L) is positive. In this example Timer1 is programmed as a timer with a prescaler factor of 1.

The following is the assembler code:

```
            List       p = 16F873
            include "P16F873.INC"
;
; Init_compare: Subroutine to program module CCP1 in compare
; mode
; with Timer1 reset when the result is positive. CCP1 pin not
; used.
; Timer1 programmed as timer with prescaler = 1.
Init_compare:
            clrf       T1CON        ; Timer1 as timer. Prescaler = 1.
                                    ; Stopped.
            clrf       CCP1CON      ; Reset CCP1 module.
            bsf        STATUS, RP0  ; Select bank 1.
            bcf        PIE1, TMR1IE ; Disable Timer1 interrupt.
            bcf        PIE1, CCP1IE ; Disable CCP1 module interrupt.
            bcf        STATUS, RP0  ; Select bank 0.
            clrf       PIR1         ; Set interrupt flags to 0.
            movlw      0x0B         ; Select comparator mode with
                                    ; Timer1 reset.
            movwf      CCP1CON ;
            bsf        T1CON, TMR1ON ; Start Timer1 count.
            return
;
; Compare: This subroutine waits for the comparison in the CCP1
; module, programmed
; as comparator is positive. This mode compares the CCPR1
; (CCPR1H and CCPR1L)
; and TMR1 (TMR1H and TMR1L) registers. When CCPR1 = TMR1, the
; result is positive
; and the subroutine ends.
;          Inputs: In CCPR1H and CCPR1L, count module of Timer1.
Compare:
            btfss      PIR1, CCP1IF ; CCPR1 = TMR1?
            goto       Compare      ; No - Wait.
            bcf        PIR1, CCP1IF ; Yes - Set comparison flag to 0.
            return                  ; Return.
            end
```

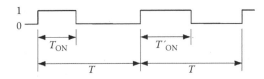

FIGURE 6.13
PWM signal. The width of the pulses is variable (T_{ON}, T'_{ON}), but the period T is constant.

6.2.3 PWM Mode

When working in pulse width modulation (PWM) mode, each CCP module and Timer2 make up a pulse width modulator whose output is in pin CCPx for the module. A PWM signal is a train of pulses with variable T_{ON} and fixed period T, as seen in Figure 6.13.

A train of pulses can be characterized by its duty cycle defined as the ratio between the time the signal is active and its period:

$$\text{Duty cycle (\%)} = \frac{T_{ON}}{T} \times 100$$

$$(6.12)$$

For example, if the pulses are active for half of a period, the duty cycle is 50%; if the pulses are active during the whole period, the duty cycle is 100%. In a PWM signal, its duty cycle changes depending on the modulating signal. Another parameter normally used to characterize a PWM signal is its *resolution*. In digital systems, time is a discrete variable causing T_{ON} to not change continuously but in different discrete time steps, ΔT_{ON}. If this interval is a submultiple of the signal period, the time that the pulses are active, T_{ON}, can only take the following discrete values:

$$T_{ON} = 0, \Delta T_{ON}, 2\Delta T_{ON}, 3\Delta T_{ON}, \ldots, T. \qquad (6.13)$$

The value R, which is the number of times that the period T can be divided, defines the resolution of the PWM signal:

$$R = \frac{T}{\Delta T_{ON}}. \qquad (6.14)$$

If R is a power of 2, that is, $R = 2^r$, r is the resolution expressed in bits:

$$r = \frac{\lg R}{\lg 2}$$

$$(6.15)$$

The resolution (r) of a PWM signal gives the number of necessary bits to determine any possible duration T_{ON} for that signal. It measures the number of parts in which we can divide the PWM pulses. For example, if $R = 1024$, the PWM signal resolution is 10 bits. This means that the duration T_{ON} of the pulses can change at time increments of T/R.

In a CCP module programmed as a PWM modulator, the period T of the PWM signal is determined by the value of the PR2 register in Timer2. For each period, T_{ON} is determined by bits DCxB9:DCxB0. These bits are distributed between registers CCPRxL and CCPxCON in the following manner: The 8 most significant bits, DCxB9:DCxB2, come from the CCPRxL register. The 2 least significant bits, DCxB1:DCxB0, come from bits CCPxCON<5:6> from the CCPxCON register. Figure 6.14 shows the configuration acquired by the CCP and Timer2 modules working in PWM mode. The modulator compares the value stored in the PR2 register with the changing value in TMR2. When both values are equal, this means that a time equal to the period of the PWM signal has elapsed. When this happens, the following actions take place:

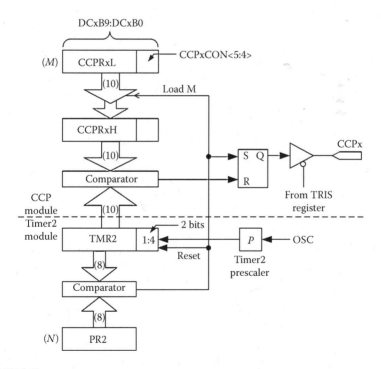

FIGURE 6.14
CCP module in PWM mode. The PWM is available at the pin CCPx. The signal period is determined by the value N stored in the PR2 register for Timer2. The pulse width is defined by the value M made up by bits DCxB9:DCxB0.

- TMR2 register is set to 0 in the next machine cycle.
- The input S for the RS bistable becomes active, making the output Q become 1. Therefore, if the CCPx pin is programmed as an output by using the appropriate bit in the TRIS register, the pin also becomes high.
- The value stored in DCxB9:DCxB0 bits is loaded in CCPRxH and in two bits internal to the module. This value is proportional to the duration T_{ON} of the PWM signal pulses.

Once these actions have taken place, a new period for the PWM signal is initiated. The value DCxB9:DCxB0, which is proportional to the duty cycle for the PWM signal, is compared against the changing value in TMR2, plus the two internal bits. When both values are equal, the input R to the bistable becomes active, making its output Q become 0. If the CCPx pin has been programmed as an output, this pin becomes low. The resulting PWM signal and associated parameters are shown in Figure 6.15.

The period (T) and the duration (T_{ON}) of the PWM signal pulses can be calculated as follows:

$$T = (N + 1) \times P \times 4 \times T_{OSC}, \tag{6.16a}$$

$$T_{ON} = M \times P \times T_{OSC}, \tag{6.16b}$$

with N being the value stored in the PR2 register, M the number made of bits DCxB9:DCxB0, P the prescaler division factor for Timer2 ($P = 1, 4, 16$), and T_{OSC} the period of the main oscillator in the microcontroller. Because M is a 10-bit number, it can be written as

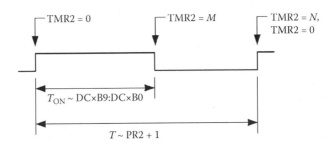

FIGURE 6.15
PWM signal generated by CCP module. Its duty cycle is determined by the value M from bits DCxB9:DCxB0 and its period is determined by the value N stored in PR2. The TMR2 register increases in each machine cycle from 0 to N. When it reaches N, it is set to 0 in the next machine cycle, thus starting a new period for the PWM signal. The signal starts its duty cycle when TMR2 is set to 0 and ends when it reaches the value M.

$$M = 4 \times M_8 + M_2, \tag{6.17}$$

with M_8 being the number stored in CCPRxL and M_2 the 2-bit number stored in CCPxCON<5:4>. To ease the change of pulse durations, it may be desirable to change only the CCPRxL register; then bits M_2 may be kept at 0. In this situation, considering Equations 6.16b and 6.17, the duration of the pulses can be found as

$$T_{ON} = 4 \times M_8 \times P \times T_{OSC}. \tag{6.18}$$

The length of the PWM pulses can be divided in M parts. Because M is a 10-bit number, it would be easy to conclude that the PWM resolution is 10 bits. However, the real resolution is lower because the highest value of M is limited by the value N stored in PR2. For this reason, the resolution depends on N. It is possible to calculate the resolution of the PWM module with the following reasoning: The smallest possible value for T_{ON} is

$$\Delta T_{ON} = P \times T_{OSC}. \tag{6.19}$$

Therefore, the number of times that this value fits into the period T is

$$R = \frac{(N+1) \times P \times 4 \times T_{OSC}}{P \times T_{OSC}} = 4 \times (N+1). \tag{6.20}$$

Because $(N + 1)$ is an n-bit number ($n = 1, 2, \ldots, 8$) and considering Equation 6.15, the resolution in bits that it is possible to achieve is

$$r = n + 2. \tag{6.21}$$

From Equation 6.18, and not using the 2 least significant bits from M, the lowest value for T_{ON} is

$$\Delta T_{ON} = \frac{T_{ON}}{M_8} = 4 \times P \times T_{OSC}. \tag{6.22}$$

In this case, the resolution is

$$R = \frac{(N+1) \times P \times 4 \times T_{OSC}}{4 \times P \times T_{OSC}} = N+1. \tag{6.23}$$

Expressing the resolution in bits results in

$$r = n. \tag{6.24}$$

Example 6.6

Program the CCP1 module in a PIC16F873 to generate a PWM signal. The PWM signal must have a constant period of 1 ms and variable duty cycle. The frequency of the main oscillator is 4 MHz ($T_{OSC} = 0.25$ μs).

The period T for the PWM signal depends on the value N stored in the PR2 register. The width of the pulses T_{ON} depends on the value M from the 10 bits distributed between the 8 bits from the CCPR1L register (these make up the 8 most significant bits of $M - M_8$) and the 2 bits from CCP1CON<5:4> that make up the 2 least significant bits of $M - M_2$. This example uses only the register CCPR1L to set the width of T_{ON}. Therefore, bits CCP1CON<5:4> are always kept at 0.

Considering $T_{OSC} = 0.25$ μs and $T = 1$ ms, using Equation 6.16 we find

$$(N + 1) \times P = 1000,$$

with P being the prescaler factor for Timer2, $P = 1, 4, 16$. Choosing $P = 4$, the number to store in the PR2 register is $N = 249$.

The variable width of the pulses is achieved by using a value between 0 and 249 in the CCPR1L register. The following is the code in assembler language:

```
        List  p = 16F873
        include  "P16F873.INC"
;
; Init_pwm: Subroutine to program CCP1 module in PWM mode
; in order to generate a PWM signal with a period of 1 ms and a
; duty cycle of 50%.
; The value of the duty cycle can be changed later.
Init_pwm:
        movlw    0x01           ; Program Timer2 stopped
        movwf    T2CON          ; and prescaler = 4.
        clrf     CCP1CON        ; Reset module CCP1.
        clrf     TMR2           ; Set Timer2 to 0.
        movlw    .124           ; PWM signal will have
        movwf    CCPR1L         ; duty cycle of 50%.
        bsf      STATUS, RP0    ; Select bank 1.
        movlw    .249           ; Count module for Timer2
        movwf    PR2            ; that is the period of the
                                ; PWM signal.
        bcf      PIE1, TMR2IE   ; Disable interrupt in
                                ; Timer2.
        bcf      PIE1, CCP1IE   ; Disable interrupt in module
                                ; CCP1.
        bcf      TRISC, 2       ; Select pin CCP1 as output.
        bcf      STATUS, RP0    ; Select bank 0.
        clrf     PIR1           ; Set interrupt flags to 0.
        movlw    0x0C           ; Module CCP1 in PWM mode
        movwf    CCP1CON        ; with DC1B1:DC1B0 to 0
                                ; (M2 = 0).
        bsf      T2CON, TMR2ON  ; Start counting for Timer2.
        return
;
; TON_pwm: This subroutine changes the duty cycle for the PWM
    signal
;                    to the value stored in W.
;                    Inputs: A number between 0 and 249 in W.
```

```
TON_pwm:
        movwf CCPR1L
        return
        end
```

7

Interrupts

This chapter describes the interrupts and interrupt systems in microcontrollers. It starts by describing basic concepts such as the possible sources for interrupts, the resources that the microcontroller needs for handling them, and how interrupts are serviced. This is followed by a deeper description of the interrupts in medium-end PIC microcontrollers with a special emphasis on the interrupt service subroutines. The chapter finishes by describing several applications in PIC microcontrollers that use interrupts.

7.1 Basic Concepts

7.1.1 Interrupt Requests and Associated Resources

An *interrupt request* (or more commonly just an *interrupt*) is an internal or external event that when serviced makes the microcontroller interrupt the execution of the current program and execute another program instead. Generally, as shown in Figure 7.1, once the program that services the interrupt request is finished, the microcontroller continues executing the instructions of the program that was being executed before the interrupt.

Interrupt requests are normally events asynchronous to the program being executed by the microcontroller. This means that an interrupt request can occur at any time during the execution of a program. Therefore, it is not possible to predict which instruction is being executed by the microcontroller when an interrupt is serviced. A microcontroller has several sources of interrupts, which can be external or internal. Internal interrupts can be originated in the microcontroller's input/output (I/O) modules, its memory, or its CPU. The most common internal interrupt sources are timers and other I/O modules. Events that occur in the memory, for example, writing in the data EEPROM, or events that occur in the CPU, for example a division by zero, are less common. External interrupts originate in peripherals and reach the microcontroller through one of its pins and associated ports.

Microcontrollers have resources to receive and process these interrupt requests. Generally, each device that can be an interrupt source has two

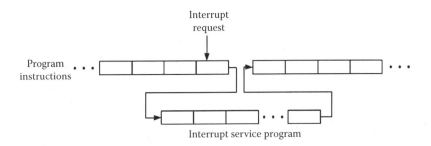

FIGURE 7.1
When an interrupt request is serviced by the microcontroller, the microcontroller finishes the instruction being executed and jumps to execute the interrupt service program. Once this finishes, it continues executing the program that was interrupted.

associated bits. These bits may be in the same register or in different registers. The function of the first bit is informative; it is a flag activated by the device that requests the interrupt. This bit can be checked by the program depending on the technique used to service the interrupt. The other bit has a control function. It is used to allow or disallow the interrupt request to reach the CPU. This is equivalent to enabling or disabling the possibility of the device to generate interrupt requests. This control bit can be manipulated by the program.

Microcontrollers also have one bit for global control of interrupts. This bit allows or disallows any interrupt request to reach the CPU. This is equivalent to enabling or disabling the whole interrupt system in the microcontroller. In order for an interrupt request to reach the CPU and be serviced, both the global interrupt system and the specific interrupt must be enabled. The control bits used to allow or disallow interrupt requests to reach the CPU are called *masks*. This is the reason for the names maskable interrupts and non-maskable interrupts. *Maskable interrupts* are those that may be enabled or disabled by software. *Non-maskable interrupts* are those that cannot be disabled by software and therefore are always enabled.

Figure 7.2 illustrates the path for all the interrupt requests to reach the CPU. Maskable interrupts have control bits associated with each interrupt source as well as the control bit for the global system. For a maskable interrupt to reach the CPU, its specific control bit and the global control bit must be set to 1. Non-maskable interrupts will always reach the CPU and will be serviced independently of the global control bit status.

When an interrupt request reaches the CPU and is serviced, the interrupt system is disabled (the global control bit is set to 0). Any new interrupt request will not reach the CPU. To be able to service new interrupt requests, the programmer must enable the system again. This is normally done by the same program that services the interrupt.

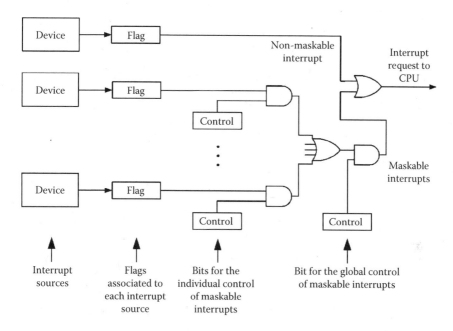

FIGURE 7.2

Path followed by interrupt requests in a microcontroller. Each interrupt source has a flag that is active with the interrupt request. Maskable interrupts reach the CPU if both the individual and global flags are enabled. This can be done by the individual and global control bits. Non-maskable interrupts always reach the CPU.

7.1.2 Servicing Interrupt Requests

Servicing an interrupt request means to interrupt the execution of the current program and to move it to execute another program, as shown in Figure 7.1. When the second program ends, it is necessary to continue with the program that was interrupted. The interrupt request that reaches the CPU is serviced after the CPU finishes the instruction it was executing. Because it is not possible to predict when an interrupt request will take place, it is necessary to remember the address for the next instruction so the program can continue once the interrupt service ends. This is done by storing the program counter (PC) in the stack, in a similar way to how it is stored in subroutine calls. For this reason, it is convenient that the service interrupt program is structured in the same way that subroutines are structured. The return instruction that finishes the execution of this subroutine allows a return to the program that was interrupted. Therefore, the interrupt service program is a subroutine that is "called" through an interrupt. This is equivalent to saying that an interrupt request is the same as inserting a subroutine call in an unpredictable location in the program.

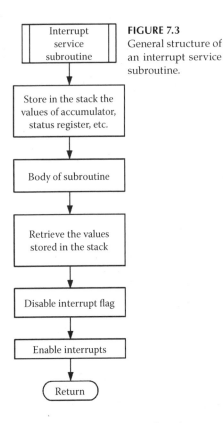

FIGURE 7.3
General structure of an interrupt service subroutine.

In general, the structure for the interrupt service subroutine is no different than that for standard subroutines. However, because it is not possible to predict at what point in the program the interrupt request will take place, the programmer needs to take some additional precautions when writing the interrupt service subroutine. The execution of this subroutine must leave unchanged the values of registers and bits with which the program is working. For example, the values of the accumulator (or the W register), the status of similar registers, as well as the arithmetic indicators cannot be altered due to the execution of the interrupt service subroutine. To keep the integrity of these registers, it is necessary to store their current values at the beginning of the subroutine. Before returning to the program that was interrupted, these values are retrieved and restored in the registers. It is also necessary to disable the interrupt flag (setting it to 0) because, in general, this flag is not automatically disabled, and then enable the interrupt system in order to return to the program that was interrupted. Figure 7.3 shows the block diagram for a generic interrupt service subroutine.

The steps to service an interrupt request are:

1. The microcontroller completes the execution of the current instruction.
2. The PC value is stored in the stack in order to remember the address of the instruction to be executed after finishing the interrupt service program.
3. The address of the interrupt service subroutine is stored in the PC, making the program branch to that address and start the execution of the subroutine.
4. The interrupt service subroutine is executed. As in any other subroutine, it ends with a return instruction.
5. When executing the return instruction, the microcontroller continues executing the program that was interrupted.

7.1.3 Fixed and Vectored Interrupts

There are two possible ways to provide the CPU with the address for the interrupt service subroutine:

1. The interrupt service subroutine is stored in a fixed memory location that the CPU already knows.
2. At the time of the interrupt being requested, the interrupt service subroutine address is given to the CPU.

The first solution is known as a *fixed interrupt* request. In this type of interrupt, the microcontroller always jumps to a fixed location in memory. This address has the first instruction in the interrupt service subroutine. Depending on the type of microcontroller, there can be a different memory address assigned to each interrupt source, or as happens with PIC microcontrollers, there is a single address for all interrupts. In any case, these addresses are always fixed. Because of its simplicity, this type of interrupt is widely used in microcontrollers.

The second solution is more flexible but it is also more difficult to implement. When the interrupt is requested, the address for the interrupt service subroutine or other information to find it is given to the CPU. This information given to the CPU is known as an *interrupt vector*, thus giving the name of this technique: *vectored interrupts*. When using vectored interrupts, the interrupt service subroutine may be located at any address in the program memory. The interrupt vector can have a different structure: the simplest structure is the address of the subroutine. It can also be a number that acts as a pointer to the subroutine (with the vector, the CPU looks for the subroutine address from a table of addresses located in memory). Vectored interrupts are widely used in microprocessors but not in microcontrollers.

Example 7.1

Interrupt system in MCS51 microcontrollers. A typical microcontroller from the 8051 family has five interrupt sources, as shown in Figure 7.4: two external interrupts, two interrupts from timers, and one interrupt from the serial port. All of them are maskable interrupts.

Each interrupt has an associated indicator (serial port interrupt has two of them) that is set to 1 if the interrupt is requested. These indicators are stored in the special function registers TCON and SCON in the microcontroller. Each interrupt source can be individually and globally enabled using control bits in the microcontroller register IE.

The interrupts in the MCS51 microcontroller family are fixed. Each interrupt source is assigned a different address to store the interrupt service subroutine. Table 7.1 shows the addresses associated with every interrupt.

This approach allows prioritizing of interrupt service. Each interrupt source can have two priority levels: low or high. A high priority interrupt can interrupt a low priority interrupt but a low priority interrupt cannot interrupt a high priority interrupt. Interrupts with the same priority level are serviced using an internal priority scale that from high to low is: external interrupt 0, timer 0, external interrupt 1, timer 1, and serial port.

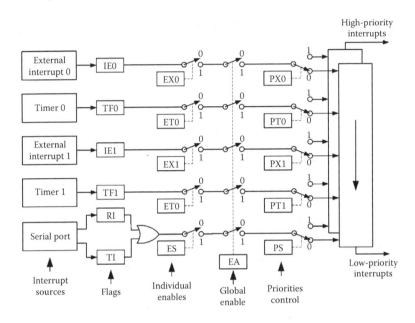

FIGURE 7.4
Interrupt system structure for a typical microcontroller in the MCS51 family.

TABLE 7.1

Associated Addresses for Every Interrupt Source in a Microcontroller from the MCS51 Family

Interrupt Source	Address
External interrupt 0	0003h
Timer 0	000Bh
External interrupt 1	0013h
Timer 1	001Bh
Serial port	0023h

Note: The addresses contain the first instruction of the appropriate service interrupt subroutine.

7.2 Interrupts in PIC Microcontrollers

7.2.1 Interrupt Sources and Associated Registers

Interrupts in medium-end PIC microcontrollers are fixed, maskable interrupts. This means that interrupts can be globally enabled or disabled, as well as each interrupt source can be individually enabled or disabled, as shown in Figure 7.7. For an interrupt request to be serviced, both the individual and global interrupts must be enabled. Global interrupts are enabled by bit GIE (global interrupt enable) from the INTCON special function register. The devices that can generate interrupt requests are enabled or disabled by bits in the INTCON, PIE1, and PIE2 registers.

Because of fixed interrupts, when an interrupt is serviced, the microcontroller executes the instruction stored in address 4 in program memory. When writing the interrupt service subroutine, the programmer must first find out the source that requested the interrupt. This is done by reading the appropriate bits in the special function registers (INTCON, PIR1, and PIR2) associated with the interrupt system in PICs.

When an interrupt is requested (assuming the global system and the specific interrupt source interrupts are enabled), the microcontroller finishes executing the current instruction, stores the value of the program counter in the stack, and jumps to address 4 in program memory. The time elapsed between the moment the interrupt is requested and the execution of the first instruction in the interrupt service subroutine located in address 4 ranges from 3 to 3.75 machine cycles. This time is called *latency time*. The exact latency time depends on the specific moment within a machine cycle in which the interrupt is requested and whether the request was internal or external. Figure 7.5 illustrates the operations carried out by the microcontroller during the latency time. First, it must finish with the instruction being executed when the interrupt was requested. Then, the value of the program counter (PC) pointing to the next instruction is stored in the stack. Finally, the microcontroller sets the PC to address 0004, making the program jump to execute the first address for the interrupt service program. When an interrupt request is being serviced, the global interrupt system is disabled (bit GIE set to 0).

Each I/O module can generate at least one interrupt request. The possible interrupt sources are:

- External interrupt in pin INT in the microcontroller
- Interrupt due to a logic change of state in inputs RB4:RB7 in port B
- Interrupt due to Timer0, Timer1, or Timer2 overflow
- Interrupt due to an event in the CCP module

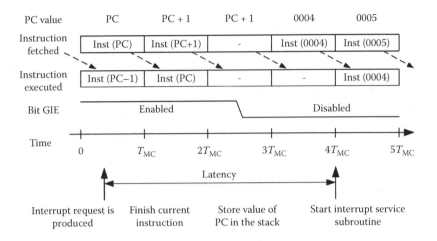

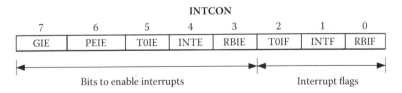

FIGURE 7.5
Microcontroller operations during latency time: the current instruction is finished, the value of the PC is stored in the stack, and the PC is loaded with 0004, causing it to execute the first instruction in the interrupt service subroutine. When this subroutine is executed, the interrupt system is disabled. The instruction retfie, used to return to the interrupted program, enables the interrupt system again.

- Interrupt in the USART serial port
- Interrupt due to the A/D converter

All medium-end PIC microcontrollers use at least one special function register to control interrupts: the INTCON register. This register controls the external interrupts (from pin INT), the interrupt due to changes in pins RB4 to RB7, and the interrupt due to Timer0 overflow. It also has a bit (GIE) to enable interrupts globally. The remaining interrupt sources are controlled with registers PIE1, PIR1, PIE2, and PIR2. Figure 7.6 shows the bits in the interrupt control register INTCON.

The details for each bit are:

GIE (global interrupt enable). This bit enables (GIE = 1) or disables (GIE = 0) the microcontroller's global interrupt system. When an

INTCON

7	6	5	4	3	2	1	0
GIE	PEIE	T0IE	INTE	RBIE	T0IF	INTF	RBIF

Bits to enable interrupts   Interrupt flags

FIGURE 7.6
INTCON register for interrupt control.

interrupt is requested, this bit is automatically set to 0. Therefore, the interrupt systems will be disabled until GIE = 1. The return instruction at the end of the interrupt service subroutine (retfie) sets bit GIE to 1, thus enabling the global interrupt system for the PIC. In a reset, bit GIE is set to 0, disabling the interrupt system.

PEIE (peripheral interrupt enable). This bit enables interrupt sources not present in the INTCON register but in the PIE registers.

T0IE, T0IF. These bits are related to Timer0 interrupt. When T0IE = 1, the Timer0 interrupt is enabled. T0IF indicates overflow in Timer0. When this bit is set to 1 it generates an interrupt request if T0IE = 1.

INTE, INTF. These bits are related to the external interrupt. Bit INTE enables this interrupt. Bit INTF is a flag that indicates the detection of an edge (raising or falling) in the input pin INT.

RBIE, FBIF. These bits are related to the interrupt due to a change of level in bits RB4 to RB7. RBIE = 1 enables this interrupt. RBIF is a flag that indicates a change in one of the pins RB4 to RB7. If RBIE = 1, an interrupt is produced.

Figure 7.7 illustrates the operation of bits in the INTCON register. The external interrupt request is done through pin INT. This pin shares functions with pin RB0 from port B in most of the medium-end PIC microcontrollers. This interrupt is produced by an edge in the signal connected to

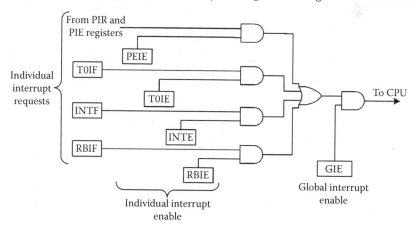

FIGURE 7.7
Role of the bits in the INTCON register in the interrupt system for medium-end PIC microcontrollers. The system is enabled or disabled with bit GIE. Each interrupt source has a control bit that enables or disables their interrupt request. Interrupt requests are shown in the flags T0IF, INTF, and RBIF independent of their progress in reaching the CPU.

pin INT. Bit INTEDG in the OPTION register (bit OPTION<6>) selects a rising or falling edge. The external interrupt is flagged by bit INTF in the INTCON register, and is enabled or disabled with bit INTE in this same register.

If pins RB4 to RB7 have been programmed as inputs and there is a change of logic level in the signals at any of these inputs, an interrupt request is generated. In this case, bit RBIF in the INTCON register is set to 1. This interrupt can be enabled or disabled with bit RBIE from the same register. This interrupt can be used to wake up the microcontroller when it is in low-power mode (sleep mode).

Figure 7.8 shows a simplified circuit used to generate the interrupt due to level changes in any of the inputs RB4 to RB7. The signal at the pin is sampled twice in two different time intervals corresponding to Q1 and Q3 in the machine cycle (Figure 2.2). If the logic level for the signal at the pin at the sampling times has changed, the output Q for the D latches will be different, making the output of the XOR gate equal to 1. This sets bit RBIF to 1, therefore generating the interrupt request. For this process to take place, the pins must be configured as input pins, that is, bit TRISB<i> = 1.

The registers PIE and PIR control the interrupts for the different peripheral modules in the PIC microcontroller. The PIE registers (PIE1, PIE2) contain the bits to enable or disable interrupts from peripherals. The PIR registers (PIR1, PIR2) contain the bits that indicate an interrupt was requested by the different peripherals. The structure of these registers, that is, the meaning of the bits and their position within a register, change from one PIC to another.

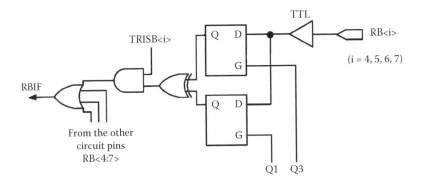

FIGURE 7.8

Simplified circuit associated to pins RB4 to RB7 programmed as inputs. The figure illustrates the generation of interrupts due to changes in the logic level at the inputs. Q1 and Q3 are internal signals corresponding to the states with their same name in a machine cycle (refer to Figure 2.2).

Example 7.2

Registers PIE and PIR in the PIC16F87x. The structure and composition of the PIE and PIR registers depend on the model of PIC used because the available I/O modules also depend on each device. The family of microcontrollers PIC16F87x has two PIE registers (PIE1 and PIE2) and two PIR (PIR1 and PIR2) registers. Their structure is shown in Figure 7.9. Table 7.2 shows the name of the bits and the I/O modules associated with them.

For an interrupt request to really interrupt the program being executed, first it is necessary that the global interrupt system in the PIC is enabled. This means that bit GIE in the INTCON register must be 1. Second, it is necessary for the interrupt source to also be enabled. For example, in the case of an external interrupt, bit INTE in the INTCON register must be 1. In these conditions, the request is successful. During the execution of the interrupt, bit GIE is automatically set to 0, disabling interrupts. Therefore,

	7	6	5	4	3	2	1	0
PIE1	PSPIE	ADIE	RCIE	TXIE	SSPIE	CCP1IE	TMR2IE	TMR1IE
PIE2	-	-	-	EEIE	BCLIE	-	-	CCP2IE
PIR1	PSPIF	ADIF	RCIF	TXIF	SSPIF	CCP1IF	TMR2IF	TMR1IF
PIR2	-	-	-	EEIF	BCLIF	-	-	CCP2IF

FIGURE 7.9
PIE and PIR registers in PIC16F87x microcontrollers.

TABLE 7.2
Some of the Bits in the PIE and PIR Registers and I/O Associated with Them

Control Bit in PIE Register	Flag Bit in PIR Register	Interrupt Source
TMR1IE	TMR1IF	Timer1
TMR2IE	TMR2IF	Timer2
CCP1IE	CCP1IF	CCP1
CCP2IE	CCP2IF	CCP2
RCIE	RCIF	USART (reception)
TXIE	TXIF	USART (transmission)
ADIE	ADIF	A/D converter
PSPIE	PSPIF	PSP
SSPIE	SSPIF	SSP
BCLIE	BCLIE	SSP (I^2C bus collision)
EEIE	EEIF	Data EEPROM

new interrupts will not be serviced. The global interrupt system is enabled again when the microcontroller executes the instruction retfie to return to the program that had been interrupted. The instruction retfie sets bit GIE to 1, thus enabling the global interrupt system again. This sets the PIC microcontroller ready to service new interrupt requests.

When a reset is produced, bit GIE is set to 0 and the microcontroller will not service any interrupt request after a reset. For example, when the microcontroller is first powered up, it starts working with the interrupt system disabled. The programmer must enable interrupts by setting bit GIE to 1.

The bits that control individual interrupts (T0IE, INTE, etc.) are not modified when an interrupt is requested. The individual interrupt flags (T0IF, INTF, etc.) are automatically set to 1 to inform about the interrupt request. They may be set back to 0 by the program that services the interrupt request.

7.2.2 Interrupt Service Subroutine Structure

Interrupts are events that can occur at any time during the execution of any instruction. It is not possible to predict the instruction being executed when an interrupt request reaches the CPU. This situation forces the programmer to take precautions to keep the integrity of the registers that are used by both the interrupted program and the program that services the interrupt. The most commonly used registers are the W and the STATUS. Therefore, their current values when the interrupt is requested must be stored in memory, and once the service interrupt subroutine has finished, these values must be restored to continue with the execution of the program. In most microprocessors and microcontrollers, the values from the registers are stored in the stack. This is done by using instructions such as PUSH register to store a register in the stack and POP register to retrieve the values. The LIFO structure of the stack allows nesting several interrupt service subroutines. Thus, the interrupt service subroutine can be interrupted by a second one even if the first one was not finished yet. This originates very complex and powerful interrupt systems.

Medium-end PIC microcontrollers do not have instructions like PUSH or POP, and the stack can only store the program counter. The absence of a stack to store the content of registers while an interrupt is being serviced makes it extremely difficult to nest interrupt service subroutines. Therefore, the interrupt system in this type of microcontroller is relatively simple. If the microcontroller is servicing an interrupt request and receives another interrupt request, the new interrupt request must wait until the first one has finished before being serviced. This is intrinsically guaranteed in these types of microcontrollers because the global interrupt system is disabled while the microcontroller services an interrupt request.

The global interrupt system is only enabled with the instruction retfie that finishes the interrupt service subroutine.

Because the stack can only store the program counter, the contents of the W and STATUS registers must be stored in the data memory or in any other available location. This can be a complex task given the bank structure for data memory and the fact that most data transfer instructions alter some bits in the STATUS register. Example 7.3 illustrates several solutions given by the manufacturer of the PIC microcontrollers.

Example 7.3

Recommended structure for an interrupt request subroutine. This subroutine uses registers TEM_W and TEMP_ST to store the contents of the W and STATUS registers. This operation seems easy to carry out with the following sequence:

```
movwf TEMP_W
movf  STATUS, W
movwf TEMP_ST
```

However, this sequence is not valid because the instruction movf affects bit Z in the STATUS register, making the value stored in the register TEMP_ST different from the original STATUS. It is necessary to use instructions that do not affect any bits in the STATUS register. One of these is the instruction swapf. Using this instruction, the segment of program becomes:

```
movwf TEMP_W
swapf STATUS, W
movwf TEMP_ST
```

This instruction stores the contents of the STATUS register in the TEMP_ST register without altering the original value of the bits. The instruction, however, swaps the nibbles in the register. Therefore, the nibbles in TEMP_ST must be swapped again before leaving the interrupt request subroutine in order to maintain the integrity of the STATUS register. The segment of program that restores the values of W and STATUS before the subroutine finishes is:

```
swapf TEMP_ST, W
movwf STATUS
swapf TEMP_W, F
swapf TEMP_W, W
```

It is also necessary to keep in mind that at least the TEMP_W register must be located in the bank that was active at the time of the interrupt. Because it is not possible to know what bank this is, there are several solutions depending on whether the microcontroller has a zone of RAM data memory that is common to all the banks. A common data memory zone is an area of memory that can be addressed from any bank with the same physical area in the data memory.

The following are the subroutines called SRAI1 and SRAI2 with the structure recommended by Microchip for interrupt request subroutines depending on whether the PIC has a common zone for data memory.

```
; This subroutine is for PICs with common RAM zone such as
; PIC16F84:
; Common RAM: An area of RAM that is the same in all banks.
; TEMP_W and TEMP_ST are defined in this common RAM zone.
;
SRAI1:
        movwf   TEMP_W      ; Store W in TEMP_W.
        swapf   STATUS, W   ; Swap the nibbles of STATUS
        movwf   TEMP_ST     ; and store result in TEMP_ST.
;
; Write here the body of this subroutine:
;
        swapf   TEMP_ST, W  ; Retrieve TEMP_ST and swap nibbles
        movwf   STATUS      ; store results in STATUS.
        swapf   TEMP_W, F   ; Retrieve TEMP_W and store it in
        swapf   TEMP_W, W   ; W without altering STATUS.
        retfie              ; Return to the interrupted program.
;
;
; This subroutine is for those PICs without common RAM, such as
; PIC16F873.
; Common RAM: An area of RAM that is the same in all banks.
; Register TEMP_W defined in any bank.
; Register TEMP_ST defined in bank 0.
;
SRAI2:
        movwf   TEMP_W      ; Store W in TEMP_W.
        swapf   STATUS, W   ; Swap nibbles in STATUS,
        bcf     STATUS, RP0 ; Select bank 0
        movwf   TEMP_ST     ; And store results in TEMP_ST.
;
; Write here body of subroutine
;
        swapf   TEMP_ST, W  ; Retrieve TEMP_ST and swap nibbles
        movwf   STATUS      ; store it in STATUS. The bank selected
                            ; is now the original where TEMP_W is.
        swapf   TEMP_W, F   ; Retrieve TEMP_W and store it in
        swapf   TEMP_W, W   ; W without altering STATUS.
        retfie              ; Return from interrupt.
;
```

Figure 7.10 illustrates the general structure for the interrupt service subroutine in PIC microcontrollers. After preserving the contents of the W and STATUS registers following one of the procedures shown in Example 7.3, the programmer must find the source that generated the interrupt request. Once the interrupt source has been identified, it can be serviced. Identifying the source can be done by reading the different bit flags to find which one is set to 1. Afterward, it is important to reset this bit to 0. Finally, the original values must be restored in the W and STATUS registers and must return to the program that was interrupted by executing the instruction retfie. This instruction enables the global interrupt system (sets GIE = 1).

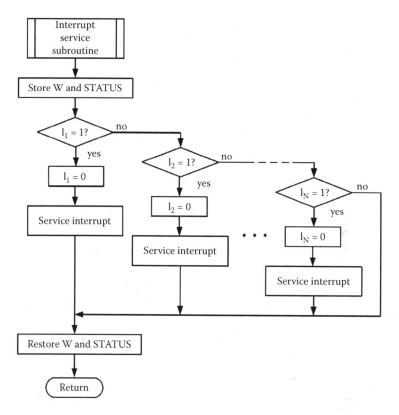

FIGURE 7.10
Interrupt service subroutine structure for medium-end PIC microcontrollers. With N potential sources, the flags I_1, I_2, ..., I_N associated to these interrupts must be consulted. These indicators are the T0IF, INTF, etc., bits from the INTCON, PIR1, and PIR2 registers.

The steps that take place in servicing an interrupt request in medium-end PIC microcontrollers are:

1. The microcontroller completes the execution of the current instruction.

2. The value of the PC is stored in the stack.

3. The PC is set to 0004, causing it to jump to this address, beginning the execution of the interrupt service subroutine.

4. The values of the W and STATUS registers are stored (see example 7.3).

5. The interrupt source is determined by checking the appropriate bit flags.

6. Once the source has been identified, its flag is set back to 0.

7. The original values for the W and STATUS registers are restored (see Example 7.3).

8. Using the instruction retfie, the microcontroller returns to the program that had been interrupted. This instruction retrieves the value of the PC from the stack and enables the global interrupt system (sets GIE = 1).

The first three steps are done automatically by the microcontroller, whereas steps 4 through 8 must be implemented within the interrupt service subroutine.

7.3 Examples of Interrupt Applications

7.3.1 Real-Time Clock

A *time base* is a set of variables whose values reflect the value of real-time in the microcontroller. For example, a time base may consist of the variables TICKS, SEC, MIN, and HOUR that count tenths of seconds, seconds, minutes, and hours. These variables are nothing more than registers in the microcontroller's data memory. The real-time clock (RTC) is a software mechanism based on periodic interrupts to actualize the time base and synchronize events. For example, with each periodic interrupt, the RTC program can increment the value of the variable TICKS and, depending on its value, actualize the rest of variables. The external events that are synchronized with the time base may be periodic. For example, periodic events can be to read one of the channels in the A/D converter every 5 s or store a value in port B every 8 s. Example 7.5 explains in further detail how these synchronizations are carried out.

The fundamental element in an RTC is the periodic interrupt that actualizes the time base. The period T of this interrupt determines the time resolution for the system. For example, if $T = 0.1$ s, the system cannot differentiate between events shorter than 0.1 s. Another important consideration is that the execution of the program that services the interrupt to the RTC must use a very short processing time, thus not limiting the ability of the microcontroller to execute other tasks.

To design an RTC we can use the interrupt from one of the timers in the microcontroller. For example, Timer0 can be programmed to periodically request an interrupt (tick of the clock) each certain number of milliseconds. This is done by using a variable (a variable that counts clock ticks) that increments or decrements with each interrupt. Other variables can be used to count seconds, minutes, hours, and so forth. The following example illustrates how to implement an RTC in the PIC16F873.

Example 7.4

Real-time clock based on a PIC16F873 with a 4 MHz crystal. The example shows how to implement an RTC using a variable time base to count ticks, seconds, minutes, and hours.

If Timer0 is programmed with a count module $N = 256$ and the prescaler with a division factor of 32, Timer0 will generate an interrupt every 8.192 ms (122.07 Hz). To reach 1 s, it is necessary to count 122 interrupts (ticks) for the counter. The register SEC implements the counter for seconds, the register MIN implements the counter for minutes, and the register HOUR implements the counter for hours.

Figure 7.11 shows the algorithm for the RTC. Although it may seem that this algorithm takes a large amount of the microcontroller's time, the reality is different because in the majority of the cases the algorithm follows the decision NO. From the 122 interrupts that occur every second, only one takes the path YES (decision block TICKS = 0?) and so on. Therefore, the algorithm uses a very short processing time. This is very desirable in an RTC. In the algorithm shown in figure 7.11, the variable TICKS decrements while SEC, MIN, and HOUR increment. The reason for this different approach is to maximize the execution speed to take the least possible amount of the microcontroller's time. Incrementing the variable TICKS and comparing its value with 122 would need more instructions than decrementing it and asking if its value is zero. Therefore, this is the fastest option.

The following is the code that implements the algorithm shown in figure 7.11:

```
;
; Real time clock
; using Timer0 interrupt
;
        list        p = 16f873
        #include <p16f873.inc>
;
; Variables of the time base:
;
TICKS       equ     0x20            ; Ticks counter.
SEC         equ     0x21        .   ; Seconds counter.
MIN         equ     0x22            ; Minutes counter.
HOUR        equ     0x23            ; Hours counter.
;
; Other variables:
;
TEMP_W      equ     0x24
TEMP_ST     equ     0x25
;
    org     0
    goto    init
    org     4
    goto    rtc
;
init:
    clrf    INTCON              ; Disable all interrupts.
    bsf     STATUS, RP0         ; Select bank 1.
    movlw   0xC4                ; Prescaler with factor 32.
    movwf   OPTION_REG          ; assigned to Timer0.
```

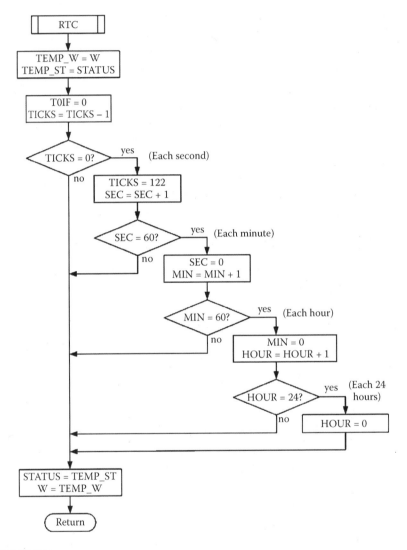

FIGURE 7.11
Block diagram with the algorithm for the real-time clock in example 7.4. The time base is made by the variables TICKS, SEC, MIN, and HOUR that count clock ticks, seconds, minutes, and hours. The algorithm is very fast because most of the times it takes the path "no" from the first decision branch. The algorithm is only fully executed (taking all the "yes" paths) for a single tick a day (just at midnight!).

```
        bcf       STATUS, RP0       ; Select bank 0.
        movlw .   0                 ; Count module = 256.
        movwf     TMR0              ; in Timer0.
        movlw     .122              ; Number of ticks per second.
        movwf     TICKS             ; in tick counter.
        clrf      SEC               ; Seconds counter to 0.
        clrf      MIN               ; Minutes counter to 0.
        clrf      HOUR              ; Hour counter to 0.
        bsf       INTCON, T0IE      ; Enable Timer0 interrupt.
        bsf       INTCON, GIE       ; Enable global interrupts.
;
prog:                               ; Main program here.
        nop
        goto      prog              ; Infinite loop.
;
rtc:
        movwf     TEMP_W            ; Store W in TEMP_W.
        swapf     STATUS, W         ; Swap STATUS nibbles,
        bcf       STATUS, RP0       ; select bank 0
        movwf     TEMP_ST           ; and store result in TEMP_ST.
                                    ;
        bcf       INTCON, T0IF      ; Set to 0 overflow flag for
                                    ; Timer0.
        decfsz    TICKS, f          ; Reached a second?
        goto      end_rtc           ; No, leave interrupt.
rtc_sec:                            ; Yes, second reached.
        movlw     .122              ; Reload variable TICKS
        movwf     TICKS             ; to initial value and
        incf      SEC, f            ; increment seconds.
        movf      SEC, W
        xorlw     .60               ; SEC = 60?
        btfsc     STATUS, Z         ;
        goto      end_rtc           ; No, return.
rtc_min:                            ; Yes, one minute reached.
        clrf      SEC               ; Set seconds to 0 and
        incf      MIN, f            ; increment minutes.
        movf      MIN, W
        xorlw .   60                ; MIN = 60?
        btfsc     STATUS, Z         ;
        goto      end_rtc           ; No, return.
rtc_hour:                           ; Yes. One hour reached.
        clrf      MIN               ; Set minutes to 0 and
        incf      HOUR, f           ; increment hours.
        movf      HOUR, W
        xorlw     .24               ; HOUR = 24?
        btfsc     STATUS, Z         ;
        goto      end_rtc           ; No, return.
rtc_day:                            ; Yes. 24 hours elapsed.
        clrf      HOUR              ; Set hours to 0.
end_rtc:
        swapf     TEMP_ST, W        ; Retrieve TEMP_ST and swap nibbles
        movwf     STATUS            ; stored in STATUS. The bank
                                    ; selected
                                    ; is now the original where
                                    ; TEMP_W is.
        swapf     TEMP_W, f         ; Retrieve TEMP_W and store it in
        swapf     TEMP_W, W         ; W without altering STATUS.
        retfie                      ; Return to interrupted program.
                                    ;
        end                         ; End of source code.
```

7.3.2 Synchronization of Events to Real-Time Clock

An RTC with a time base that counts fractions of seconds, seconds, minutes, and so forth allows us to synchronize several events with this time base. With this, each event occurs periodically at the same or different time intervals. Figure 7.12 illustrates, using a specific case, how to proceed in general to synchronize events with a time base built on an RTC. In this case, there are two events called EVENT1 and EVENT2 that must be executed periodically at 3 s and 5 s each. The procedure to achieve this is described next.

The RTC program must have as many variables as events needed to be synchronized. Each one of these variables sets the repetition time for the event. In this example, the events need to be executed each 3 s and 5 s. Therefore, the RTC program must have two variables that are incremented every second until they reach 3 s and 5 s, respectively. In Figure 7.12, these variables are called SEC3 and SEC5. The RTC program must also have one flag for each event to be synchronized. These flags are set to 1 when the time to execute the event has been reached. These flags can be bits in any of the microcontroller registers. The example shown in Figure 7.12 uses bits 0 and 1 of a register called FLAGS.

The main program periodically checks the status of these flags. If it finds any of them at 1, it means that the event must be executed. This can be done by calling the subroutine that implements the appropriate action. After the event has been executed, the flag needs to be set back to 0. From a programming point of view, these flags are global variables because they must be accessed by both the main program and the program to service the interrupt.

Example 7.5

Synchronizing two events to a time base. Using a PIC16F873 with a 4 MHz crystal, implement an RTC, a time base, and execute two events associated with the time base:

1. The value at pin RB0 must alternate between 0 and 1 every 3 s.
2. The value at pin RB1 must alternate between 0 and 1 every 5 s.

The solution to this problem follows the algorithm shown in figure 7.12. Using the 4 MHz clock, programming Timer0 with a count module of $N = 256$ and using a division factor of $P = 32$ for the prescaler, the interrupt to Timer0 will occur every 8.192 ms (122.07 Hz). To reach 1 s it is necessary to count 122 interrupts (ticks) in the counter. The program uses a register (TICKS) to count them. Registers SEC3 and SEC5 are used to implement counters from 0 to 3 s and from 0 to 5 s. To indicate that 3 s or 5 s have been reached, the program uses bit 0 and bit 1 from the FLAGS register (FLAGS<0>, FLAGS <1>). These

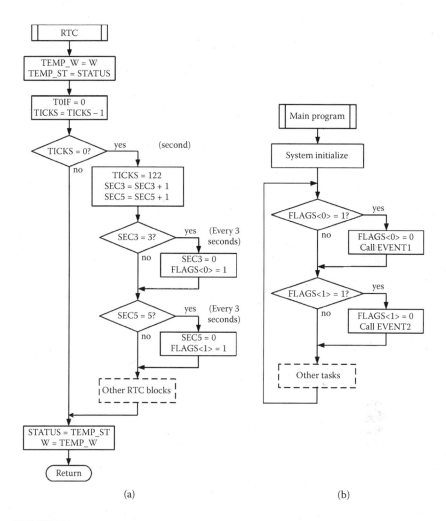

(a)   (b)

FIGURE 7.12
Synchronization of events to RTC. EVENT1 and EVENT2 are executed every 3 s and 5 s.
(a) Block diagram for the RTC with the counters SEC3 and SEC5 that increment every second until they reach 3 s and 5 s, respectively. Bits 0 and 1 in the FLAGS register indicate the occurrence of these times. (b) Block diagram for the main program that continuously checks the value of the time flags in order to execute the appropriate events. The flag bits are set to 1 by the RTC and set to 0 by the main program.

bits are set to 1 every 3 s and 5 s. The main program checks their value and sets back to 0 when the event occurs.

The program code is shown below:

```
;
; Real time clock using Timer0 interrupt.
; This time base synchronizes events EVENT1 and EVENT2,
; each 3 and 5 seconds.
```

```
        ;
        list        p = 16f873
        #include <p16f873.inc>
TICKS       equ     0x20            ; Ticks counter.
SEC3        equ     0x21            ; Counter of seconds up to 3.
SEC5        equ     0x22            ; Counter for seconds up to 5.
FLAGS       equ     0x23            ; Event flags in bits 0 and 1.
TEMP_W      equ     0x24
TEMP_ST     equ     0x25
        org         0
        goto        init
        org         4
        goto        rtc
init:
        clrf        PORTB
        clrf        INTCON          ; Disable interrupts.
        bsf         STATUS, RP0     ; Select bank 0.
        movlw       0xC4            ; Prescaler, value 32 assigned
        movwf       OPTION_REG      ; to Timer0.
        clrf        TRISB           ; Port B output port.
        bcf         STATUS, RP0     ; Select bank 0.
        movlw       .0              ; Count module = 256
        movwf       TMR0            ; in Timer0.
        movlw       .122            ; Number of ticks per second.
        movwf       TICKS           ;
        clrf        SEC3            ; SEC3 counter to 0.
        clrf        SEC5            ; SEC5 counter to 0.
        clrf        FLAGS           ; Flag events to 0.
        bsf         INTCON, T0IE    ; Enable Timer0 interrupt.
        bsf         INTCON, GIE     ; Enable global interrupt system.
prog:
        btfsc       FLAGS, 0        ; FLAGS<0> = 0?
        call        even1           ; No, carry out event 1.
        btfsc       FLAGS, 1        ; FLAGS<1> = 0?
        call        even2           ; No, carry out event 2.
        goto        prog
event1:
        bcf         FLAGS, 0        ; Set FLAGS<0> to 0
        btfsc       PORTB, 0        ;   PORTB<0> = 0?
        goto        event1_set0     ; No. It is 1. Set it to 0.
event1_set1:                        ; Yes. It is 0. Set it to 1.
        bsf         PORTB, 0
        return
event1_set0:                        ; Set to 0.
        bcf         PORTB, 0
        return
event2:
        bcf         FLAGS, 1        ; Set FLAGS<1> to 0.
        btfsc       PORTB, 1        ; PORTB<0> = 0?
        goto        event1_set0     ; No. It is 1. Set it to 0.
event2_set1:                        ; Yes. It is 0. Set it to 1
        bsf         PORTB, 1
        return
event2_set0:                        ; Set to 0.
        bcf         PORTB, 1
        return
rtc:
        movwf       TEMP_W          ; Store W in TEMP_W.
        swapf       STATUS, W       ; Swap STATUS nibbles,
        bcf         STATUS, RP0     ; select bank 0
        movwf       TEMP_ST         ; and store result in TEMP_ST.
        ;
```

```
            bcf         INTCON, T0IF  ; Clear Timer0 flag?
            decfsz      TICKS, f      ; Reached one second?
            goto        end_rtc       ; No. Leave interrupt.
                                      ;
    rtc_sec:
            movlw       .122          ; Yes. Reload value of TICKS.
            movwf       TICKS         ; with number of ticks per
second.
            incf        SEC3, f       ; Increment 3 s counter.
            incf        SEC5, f       ; Increment 5 s counter.
                                      ;
            movf        SEC3, w
            xorlw       .3            ; SEC3 = 3?
            btfsc       STATUS, Z     ;
            goto        rtc_sec1      ; No. Continue.
            clrf        SEC3          ; Yes. 3 s elapsed: Set SEC3 to 0,
            bsf         FLAGS, 0      ; set to 1 flag FLAGS<0>
                                      ; and continue.
    rtc_sec1:
            movf        SEC5, w
            xorlw       .5            ; SEC5 = 5?
            btfsc       STATUS, Z     ;
            goto        end_rtc       ; No. Continue.
            clrf        SEC5          ; Yes. 5 s elapsed: Set SEC5 to 0,
            bsf         FLAGS, 1      ; set to 1 flag FLAGS<1>
                                      ; and continue.
    end_rtc:
            swapf       TEMP_ST, W    ; Retrieve TEMP_S and swap nibbles
            movwf       STATUS        ; store in STATUS. The selected bank
                                      ; is now the original where
TEMP_W is.
            swapf       TEMP_W, F     ; Retrieve TEMP_W and store it in
            swapf       TEMP_W, W     ; W without altering STATUS.
            retfie                    ; Return to interrupted program.
            end
```

7.3.3 Protection against Hardware Malfunctions

In programs that must wait for signals coming from hardware external to the microcontroller, it is useful to build some level of protection to avoid the software falling into an infinite loop in case the communication between the microcontroller and the external hardware fails. In these situations, it is a good design idea to limit the waiting time to a reasonable value. This maximum waiting time is known as the *time-out*.

Let's assume a situation in which a peripheral needs to be serviced and the need for service comes from a signal external to the microcontroller. If the hardware were always working correctly, the algorithm shown in Figure 7.13a would be enough to service the peripheral. However, if there is an error in the external hardware, the waiting time in this algorithm would be infinity, a situation totally unacceptable.

A solution to limit this waiting time is to use a variable in the RTC that counts the waiting time that has elapsed. The peripheral service program must break the loop when this variable reaches a preset value that corresponds to the maximum waiting time decided by the programmer.

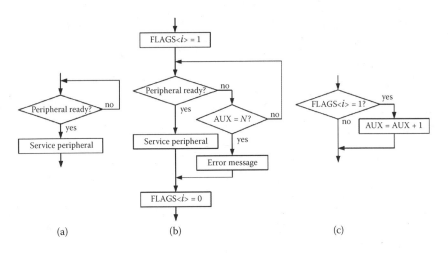

(a)   (b)   (c)

FIGURE 7.13
Protection against hardware errors when servicing peripherals. If servicing a peripheral, it is necessary to wait for an external signal. The algorithm shown in (a) has the risk of an infinite time look. The algorithm shown in (b) limits the waiting time using the variable AUX. If AUX reaches the value N that corresponds to the maximum waiting time, this means that there is a hardware error and an error message is displayed. (c) Shows the section of the RTC in which AUX is incremented. AUX must only be incremented if the program has entered the loop to wait for the peripheral. This is indicated to the RTC by bit FLAGS<i>. The place to code the section shown in (c) within the RTC depends on when the variable AUX is incremented.

Figures 7.13b and 7.13c illustrate this solution. The peripheral service program (Figure 7.13b), just before starting the waiting loop, activates a control variable (bit i in FLAGS register). This indicates to the RTC that it must start counting the waiting time. This waiting time is stored in the variable AUX. In other words, AUX counts the waiting time only if the control variable is active. The block shown in Figure 7.13c must be inserted in the RTC so AUX increments at every clock tick, or at every second, or at every unit of time decided by the programmer. When the variable AUX reaches the value N that corresponds to the maximum waiting time, this means that there is a hardware malfunction. The program interrupts the waiting loop and can send an error message informing of the error.

8

Serial Input and Output

This chapter focuses on serial input and output in microcontrollers. It starts by explaining the basic concepts regarding serial transmission of information, formats used, parameters needed, and interfaces. The chapter continues by describing the serial ports available in medium-end PIC microcontrollers and finishes by giving examples of programs for serial data transmission.

8.1 Basic Concepts

8.1.1 Introduction to Serial Data Transmission

Serial transmission of binary data consists of sending the bits of a word one by one in a consecutive form and using the same pins. For example, the 8-bit word B2h = 10110010b can be seen and transmitted as a data signal in which the bit 0 is represented by a low voltage level (V_L) and the bit 1 is represented by a high voltage level (V_H). This data signal can be generated in synchrony with a clock signal whose period determines the length of a bit, as shown in figure 8.1.

The data signal is characterized by its *transmission velocity* (v_T), which is defined as the inverse of the duration of a bit. If each bit lasts τ seconds, the transmission velocity is

$$v_T = \frac{1}{\tau} \text{ bit/s} . \tag{8.1}$$

For the transmitted information to be correctly interpreted by the receiver, there has to be some form of synchronization between transmitter and receiver. For short-distance transmission, the clock signal can be sent with the data signal in order to synchronize. This type of communication is called synchronous communication. However, sending the clock signal with the data signal is not a viable solution for long-distance transmission due to the additional costs involved in sending this additional signal that does not contain information. Although the clock signal is not present in the receiver, the receiver must know, nevertheless, the duration of each bit

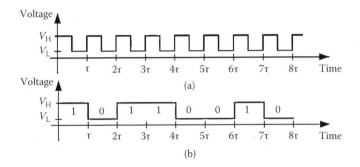

FIGURE 8.1
Serial transmission of a byte. (a) Clock signal. (b) Data signal. Each bit, represented by a high (V_H) or low voltage (V_L), is sequentially transmitted.

and the moment at which the transmission starts. Knowing the duration of each bit can be accomplished by building a clock at the same frequency in both the receiver and transmitter. Knowing when the transmission starts can be accomplished in two different ways. The first method is by marking in some way the beginning of a new word; the second method is by marking the beginning of each block of words. These two approaches are known as asynchronous communication and synchronous communication, respectively. When using asynchronous communication, the synchronization between the transmitter and receiver is done word by word. When using synchronous communication, the synchronization is done at each block of words. Both of these approaches introduce a certain amount of redundant information in the data transmitted in order to establish the necessary synchronization between transmitter and receiver.

The term *synchronous communication* is used to identify both data transmission in which the clock signal is sent with the data signal, as well as data transmission without sending the clock signal but synchronizing with each block of words. On the other hand, the term *asynchronous communication* is used only for data transmission in which the synchronization is done word by word.

The transmitter and receiver must coordinate using a *communication protocol*. This is a set of rules agreed upon by the transmitter and the receiver to ensure the correct transmission of data. There are two main types of communication protocols:

- Byte-oriented protocols in which all the transmitted words have 8 bits. An example of a byte-oriented protocol is the IBM Binary Synchronous Communication (BISYNC) Protocol.
- Bit-oriented protocols in which the blocks of transmitted data do not necessary have to have 8 bits. That is, the transmitted data can be seen as sets of bits rather than sets of bytes. Examples of

bit-oriented protocols are high-level data link control (HDLC), synchronous data link control (SDLC), and carrier sense, multiple access with collision detection (CSMA/CD), widely used in local computer networks that follow the IEEE 802.3 protocol: *Ethernet Network Standard.*

8.1.2 Asynchronous Communication

Asynchronous communication is characterized by introducing a synchronization element in each transmitted data. This synchronization element consists of an additional bit with the value 0 to indicate the beginning of each word and a bit with the value 1 to indicate the end of the word. The initial 0 bit is called the *start* or *space* bit, while the final 1 bit is called the *stop* or *mark* bit. When the transmitter pauses because it does not have words to transmit, it keeps a sequence of stop bits in its output; that is, it keeps the output at 1. Figure 8.2 shows the format of an asynchronous signal where it can be seen how the synchronization occurs with each data transmitted.

8.1.3 Synchronous Communication

Synchronous communication without transmitting the clock signal means data synchronization is at blocks of words instead of the individual word synchronization used by asynchronous communication. Here, in order to start the transmission of a data block, the transmitter introduces a synchronization element that can be a unique word or a unique bit pattern. When the transmitter finishes sending the data block and does not have more data to send, there is a pause in which the transmitter needs to keep the line in a predetermined state, generally at 1. Figure 8.3 illustrates the general signal format for synchronous communication.

The synchronization element, commonly called FLAG, is a unique sequence of bits that is not repeated during the data transmission. This element must start with a 0 bit for the receiver to determine that the pause has ended. The word 7Eh (01111110), containing a sequence of 6 bits at 1, is commonly used for this purpose. To differentiate this control sequence

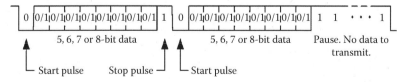

FIGURE 8.2
Signal format for asynchronous transmission. Start pulse always lasts 1 bit. Stop pulse can last 1, 1½, or 2 bits.

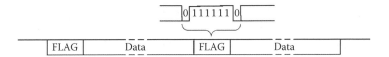

FIGURE 8.3
Signal format for synchronous transmission. Synchronization occurs at the beginning of each data block. Each block consists of a sequence of words (bytes) or simply a set of N bits.

from data with the same bits, the transmitter adds an additional 0 to any sequence of 5 bits at 1. At the end, the receiver removes the additional 0, thus recovering the initial bit sequence.

When using asynchronous communication in which the synchronization is done character by character using the start and stop bits, it is necessary to transmit 10 bits for each 8 bits of information. Therefore, 20% of the transmission time is lost by the synchronization. Synchronous communication uses its communication time more efficiently because, after the transmitter and receiver are synchronized, they only transmit/receive data.

8.1.4 Connection between Equipment: RS-232C Interface

Establishing long-distance communication requires the participation of several pieces of equipment:

- Data terminal equipment (DTE). This is the equipment that produces or receives the data signal.
- Data communication equipment (DCE). This is the equipment used to condition the data signal to the transmission medium, or the equipment that receives this data signal from the transmission medium and prepares it for the receiver.

The personal computer is a very common DTE that can generate an asynchronous data signal. When this signal needs to be transmitted to another computer using a telephone channel as the transmission media, it is then necessary to use an additional piece of equipment (DCE) to condition the data signal to the telephone channel. The reverse process needs to be done on the receiver side. This DCE is called a modem (modulator–demodulator). Figure 8.4 shows the basic block diagram for this form of data communication.

The connection between DTE and DCE has been normalized since the 1960s by the CCITT that now is part of the International Telecommunication Union (ITU). One of the most common standards for asynchronous communication using low speeds is known as RS-232C (Recommended Standard 232, Revision C). This standard was first developed by the Electronics Industries Alliance (EIA) to connect data equipment short

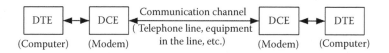

FIGURE 8.4
Simplified block diagram representing a communication system. The data terminal equipment (DTE) can be a computer and the data communication equipment (DCE) can be a modem. The communication channel can have other DCEs.

distances in a noisy environment. This interface is so common that until very recently all personal computers had a serial RS-232C interface and its connector. Since approximately 2004, personal computers with a single serial port now use a universal serial bus (USB) interface.

The signals involved in the RS-232C standard use negative logic:

- Logic level 0: Voltage up to +25 V with no load. Voltage between +3 V and + 15 V with load.

- Logic level 1: Voltage up to –25 V with no load. Voltage between –3V and –15 V with load.

Table 8.1 shows the most commonly used signals used in the RS-232C interface.

The connection of equipment using the RS-232C interface uses the connections shown in Figure 8.5. The connection between a DTE (e.g., a personal computer) and a DCE (e.g., a modem) follows the configuration shown in Figure 8.5a. The scheme shown in Figure 8.5b is used to connect two DTEs to each other, for example, a computer with another computer or a printer. It is important to realize how these connections differ from each other. The connection shown in Figure 8.5b is called a *null modem*,

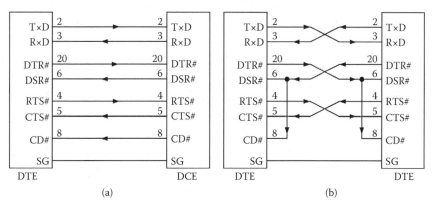

FIGURE 8.5
Equipment connection using the RS-232C interface. (a) Connection between a DTE and a DCE. (b) Null modem connection between two DTEs.

TABLE 8.1

Signals Involved in the RS-232C Interface

Connector		Name of Signal	Direction of Signal	
25D	9D		From DCE	To DCE
1		Protective ground (GND)		
2	3	Transmitted data (TxD)		X
3	2	Received data (RxD)	X	
4	7	Request to send (RTS)		X
5	8	Clear to send (CTS)	X	
6	6	Data set ready (DSR)	X	
7	5	Signal ground (SG)		
8	1	Rcvd line signal detect (data carrier detect: DCD)	X	
9		—		
10		—		
11		Select standby		X
12		—		
13		—		
14		—		
15		Transmit signal element timing	X	
16		—		
17		Receiver signal element timing	X	
18		Test	X	
19		—		
20	4	Data terminal ready (DTR)		X
21		—		
22	9	Ring indicator (RI)	X	
23		Speed select		X
24		—		
25		—		

Note: Data signals are RxD and TxD. The rest of the signals have control functions.

although this name is commonly used for any configuration different from the one shown in figure 8.5a.

8.1.5 The I²C Bus

The inter-integrated circuit (I²C) bus was developed by Philips to interconnect integrated circuits on the same printed circuit board using only three lines. This bus has become a standard for the interconnection and synchronous serial data communication between different devices located at short distances: microcontrollers, memories, D/A and A/D converters, and so forth. In addition to the ground line, this bus only uses two additional

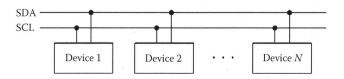

FIGURE 8.6
Device connection using the I²C bus. SDA is the data line and SCL is the clock line. The devices connected can be microcontrollers, memories, displays, and so forth. Each device is identified by its address. In any given moment, one of the devices acts as a master and the others as slaves. The clock signal is generated by the master, which can be transmitter or receiver. The data signal is generated by the transmitter, which can be a master or a slave. The figure does not show the ground connection for each device.

lines: one for data transfer, serial data line (SDA); and one for the clock signal, serial clock line (SCL). This bus can achieve transmission velocities of up to 100 kbits/s in its low-speed mode, 400 kbits/s in fast mode, and 3.4 Mbits/s in high-speed mode. Figure 8.6 shows several devices connected using the I²C bus.

One of the devices in this type of communication acts as a master, while the rest of the devices act as slaves. Both masters and slaves can be transmitters or receivers. The device acting as master initiates the communication, generates the clock signal, and ends the communication. The I²C bus is a multimaster bus meaning that there can be several masters connected to the bus, although there is a single master at any given moment. Each device has a unique address that identifies it during the communication. The addresses can have 7 or 10 bits.

Figure 8.7 illustrates the circuits that connect the devices to the I²C bus. Output circuits are open-drain or open-collector. Each bus line has

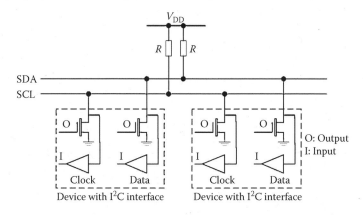

FIGURE 8.7
Inputs and outputs for devices connected to an I²C bus. Each line bus is bidirectional. The open-drain outputs and the pull-up resistors R allow a wired AND in the bus. When an output transistor saturates, it sets the line to 0.

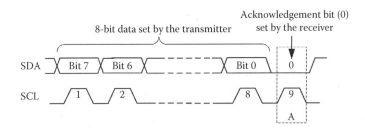

FIGURE 8.8
Data and address transfer in the I²C is organized in 8-bit words. Each word sent by the transmitter has an acknowledgment bit set by the receiver in the SDA line during the ninth clock pulse in the SCL line. Each bit in SDA must be stable when the clock pulse is 1 and can only change when the clock is at 0.

a pull-up resistance to create the logic function AND among all the lines connected to the bus. For a bus line to be set to 1, all output transistors must be off. However, when one single output transistor saturates, it sets the line to 0. This makes this device dominate the bus line. The I²C bus is bidirectional, meaning that each line can be an input or an output.

The SDA lines in the bus transfer data and the device addresses. All the information is organized in 8-bit words. Each time that the transmission of a byte through the SDA line is completed, the receiver must respond with an acknowledgment bit (A). This bit is a 0 set by the same SDA line during the following clock pulse in SCL, as shown in figure 8.8. To allow the receiver to insert the acknowledgment bit A, the transmitter frees the SDA line after transmitting the last bit and waits for the bit A = 0 in SDA before continuing with the transmission of a new byte.

As seen in figure 8.8, each SDA bit is transferred synchronized with the clock signal SCL. When the clock signal is at 1, the bit in the SDA line must be in a stable state. Changes in the SDA line can only occur when the SCL clock signal is at 0.

The communication between two or more devices in the I²C bus is always initiated and finished by the device acting as master. The communication begins when the master device generates the starting condition and ends when it generates the condition for stopping. Both conditions are identified by data transition in the SDA line while the clock signal is at 1, as shown in figure 8.9. The starting condition is identified by a transition from 1 to 0 in the SDA line when SCL is at 1. The stopping condition is identified by a transition from 0 to 1 in the SDA line when SCL is at 1.

Once the master generates the starting condition, it stores in the SDA line the address of the slave with which it desires to establish communication. From this moment, the master indicates that it will receive or transmit data with the bit R/W# that is transmitted at the end of the address. When R/W# = 0, the master is a transmitter; when R/W# = 1, the master is

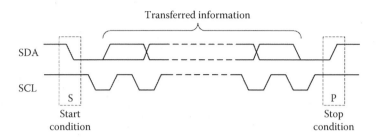

FIGURE 8.9
Start (S) and stop (P) conditions used by the master device to initialize and finish communication.

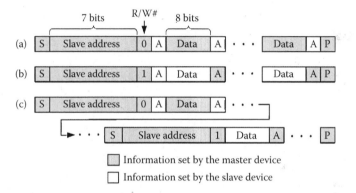

FIGURE 8.10
Communication format in three possible situations. (a) The master transmitter writes data in the slave receiver. (b) The master receiver receives data from the slave transmitter. (c) The master is initially a transmitter (R/W# = 0) and then becomes a master receiver, repeating the starting condition and sending the slave address by setting bit R/W# to 1.

a receiver. Data transfer starts after this information is sent, always with the acknowledgment bit from the receiver. Figure 8.10 illustrates the communication format using the I²C bus in three possible scenarios: (a) the master acting as transmitter sends data to the slave acting as a receiver; (b) the master acting as a receiver receives data from a slave transmitter; and (c) a master that was initially a transmitter becomes a receiver. To move from transmitter to receiver, the master repeats the starting condition (S) and sends the slave address setting the bit R/W# to the appropriate value. Independently of the master being transmitter or receiver, the address is always set by the master. It is also the master that always initiates and terminates the communication.

The initial version of the I²C bus used 7-bit addresses allowing the connection of up to 128 devices. Further versions of the bus use 10-bit addresses, allowing a higher number of connected devices. These new

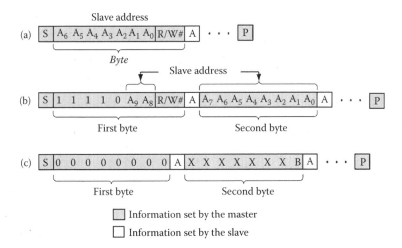

FIGURE 8.11
Addresses of devices connected to the I²C bus can be: (a) 7 bits or (b) 10 bits. (c) General call format.

versions can also understand 7-bit addresses. Figure 8.11 illustrates the general format of these addresses.

In general, 7-bit addresses are given in a single byte, except when the master carries out a general call. A general call is a call of attention to all the devices of the bus. It is followed by some information that specifies the objective of the call. Two bytes are used. The master first emits a byte "0" followed by a second byte that specifies the objective of the call. The actions that can be carried out are classified according to the value of the least significant bit of the second byte (bit B).

As shown in figure 8.11, not all possible addresses are available to be used by the devices connected to the I²C bus. Seven-bit addresses 78h to 7Bh (in binary: 11110XX) cannot be used to identify any device because these represent the first byte in a 10-bit address. There are also several addresses reserved for further developments. These details for the I²C bus are described in the specifications published by Philips (www.nxp.com).

8.2 The USART Serial Port in PIC Microcontrollers

Medium-end PIC microcontrollers have a serial communications port called the universal synchronous asynchronous receiver transmitter (USART) or serial communication interface (SCI). This port can be configured to establish a simultaneous asynchronous bidirectional communication (full duplex) or nonsimultaneous synchronous (transmitting the clock signal) bidirectional communication (half duplex).

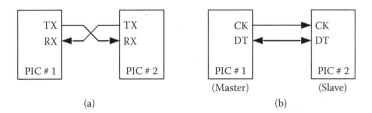

FIGURE 8.12

Pins TX/CK and RX/DT used by the USART serial port. (a) Connection between two PICs in asynchronous mode. TX and RX are data pins that transmit and receive, respectively. The clock signal is not transmitted. (b) Synchronous mode connection. CK is the clock pin: output in the master device and input in the slave device. DT is the data pin: output in the transmitter and input in the receiver.

8.2.1 General Description

The USART serial port uses the TX/CK and RX/DT pins from the micro-controller, which normally share functions with two pins in the parallel port C. In asynchronous mode, TX/CK is the pin that the transmitter uses to transmit data and RX/DT is the receiver pin for data input. In synchronous mode, TX/CK is the pin to output the clock signal if the device has been configured as a master, or the pin for the input clock signal if it has been configured as a slave. RX/DT is the bidirectional pin for data signal. Figure 8.12 shows the use of these terminals between two PIC microcontrollers.

The USART serial port uses two special function registers, TXREG and RCREG, which are used to store the data to be transmitted or the data received. The USART serial port also uses the registers TXSTA and RCSTA to control the port, and the register SPBRG to establish the communication speed. It also uses some bits from the PIE and PIR registers for control purposes and to notify of interrupt requests.

8.2.2 Asynchronous Mode

The USART serial port in asynchronous mode allows simultaneous bidirectional data communication (full duplex). This means that during the communication between two USART devices in asynchronous mode (figure 8.12a), each device can transmit and receive data at the same time. The transmitted or received signal consists of 8 bits preceded by the start bit (0) and followed by the stop bit (1). It is also possible to program the port to transmit or receive a ninth data bit as shown in Figure 8.13b.

The serial port has special function registers to manipulate the data to be transmitted or received. These are the TXREG and RCREG registers. TXREG is the register to store the data to be transmitted through the TX pin. RCREG is the register that stores the data received by the RX pin.

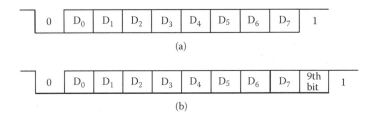

(a)

(b)

FIGURE 8.13
Format of the asynchronous signal in the serial port in medium-end PICs. (a) Data signal.
(b) Data signal with the additional ninth bit.

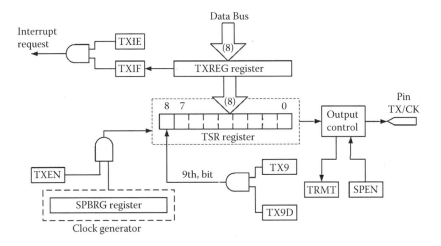

FIGURE 8.14
Transmitter in the USART serial port.

These registers do not include the ninth bit if it exists. Figure 8.14 shows the schematic of the USART serial port transmitter. In asynchronous mode and transmission, once the TXREG is empty because the data has been moved to the TSR shift register as it is being transmitted, the bit TXIF in the PIR register is set to 1. This indicates that the port is ready to transmit new data that must be stored in TXREG. If the transmission interrupt is enabled (TXIE = 1 in PIE register), when TXIF = 1 it generates an interrupt request. The bit TXIF is automatically set to 0 when new data is loaded in TXREG.

Figure 8.15 shows the schematic of the USART serial port receiver. In asynchronous mode and reception, the data received by the pin RX is sampled at a frequency 64 times higher than the clock frequency and is temporarily placed in the RSR shift register before being stored in RCREG. The RCREG register may be read by the program. In reality, there are two registers to store data organized with a FIFO (first in, first out) structure. This means that while data is being received in the RSR register, there

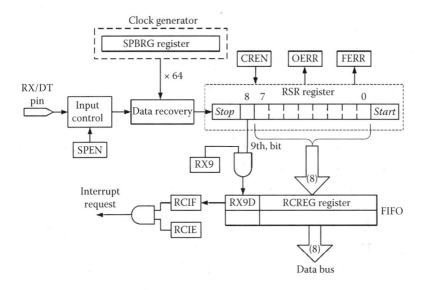

FIGURE 8.15
Receiver in the USART serial port.

TXSTA

7	6	5	4	3	2	1	0
CSRC	TX9	TXEN	SYNC	-	BRGH	TRMT	TX9D

RCSTA

7	6	5	4	3	2	1	0
SPEN	RX9	SREN	CREN	-	FERR	OERR	RX9D

FIGURE 8.16
TXSTA and RCSTA registers used to control data transmission (TXSTA) and reception (RCSTA) in the USART serial port.

can be two additional data in the RCREG register waiting to be read by the program. When the RCREG register contains data, the bit RCIF in the PIR register is set to 1. This indicates that data has been received. If the bit RCIE in the PIE register is set to 1, it generates an interrupt request. The bit RCIF is set to 0 when the RCREG register does not contain data.

The control of transmission and reception is established with the TXSTA and RCSTA registers, shown in Figure 8.16. The TXSTA register allows for: selecting the transmission mode as synchronous (bit SYNC = 1) or asynchronous (bit SYNC = 0); enabling transmission (bit TXEN = 1); enabling the transmission of the ninth bit (bit TX9 = 1) and storing its value in bit TX9D; and selecting the transmission speed in asynchronous mode with the bit BRGH (high baud rate select bit) as high speed (BRGH = 1) or low speed (BRGH = 0). The status of the shift register TSR in the transmitter can be known by the bit TRMT (transmit shift register status bit). TRMT = 1 means that the TSR register is empty. The bit CSRC (clock source select bit) is used to program the USART serial port as a master (CSRC = 1) or as a slave (CSRC = 0) for synchronous communication. It does not have any effect in asynchronous mode.

The RCSTA register allows for: enabling the serial port with the bit SPEN (serial port enable bit) set to 1, meaning that the RX and TX pins are configured as serial port terminals; enabling data reception with the bit CREN (continuous receive enable bit) set to 1; and enabling the reception of the ninth bit with RX9 = 1, storing its value in RX9D. The bit FERR (framing error bit) indicates the reception of nonvalid data (stop bit is not 0) when its value is 1. The bit OERR (overrun error bit) is set to 1 if it has stopped receiving data. The bit SREN (single receive enable bit) enables the reception of a single data in synchronous mode but does not have any effect in asynchronous mode.

8.2.3 Synchronous Mode

The USART serial port in synchronous mode is able to: (a) establish non-simultaneous bidirectional data communication (half duplex); (b) simultaneously transmit or receive data and clock signals; and (c) communicate between two devices using a master–slave approach.

In synchronous mode, the communication is bidirectional but not simultaneous (half duplex). Each device can transmit and receive data but these two operations cannot occur simultaneously; when the serial port transmits, it cannot receive and when it is receiving, it cannot transmit. Transmitters and receivers follow the schematics shown in Figures 8.14 and 8.15. In this communication mode, the clock signal is available at one of the microcontroller pins (TX/CK) and the data signal in pin TX/DT. The data signal consists of the sequence of bits from the words to be transmitted. Figure 8.17 shows the typical data and clock signals handled by the USART serial port in synchronous communication. It can be seen that they do not contain the start and stop bits that are typical in asynchronous communication.

The USART serial port in synchronous mode can act as a master or slave. The master is the device that generates the clock signal either as a

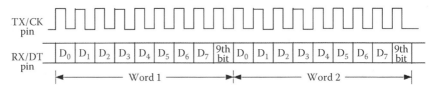

FIGURE 8.17
Clock and data signals in the synchronous transmission through the USART serial port. The transmission of the ninth bit is optional.

transmitter or as a receiver. The master initiates and ends the communication with the slave by introducing or removing the clock signal. The TX/CK pin can be used as input or output for the clock signal; it is an output for the master and an input for the slave. Figure 8.12b illustrates the connection between two microcontrollers using the pins in the USART serial port in synchronous mode. An important feature is that a slave can receive or transmit data even when it is in low-power mode (sleep). In this case, the PIC wakes up and can generate an interrupt request if the global enable bit for the interrupts allows it (GIE = 1).

In synchronous mode, the USART serial port is programmed by setting bit SYNC in the TXSTA register to 1. When bit SPEN in the RCSTA register is set to 1, the pins in the USART serial port are enabled: CK for the clock and DT for the data signal. In transmission, once the TXREG is empty because the data has been moved to the shift register TSR, the bit TXIF in the PIR register is set to 1. This indicates that the port is ready to transmit new data. If the transmission interrupt is enabled (TXIE = 1 in the PIE register) it generates an interrupt when bit TXIF is set to 1. The bit TXIF is automatically set to 0 when new data is loaded in TXREG. In reception mode, when there are data in RCREG, the bit RCIF in the PIR register is set to 1. If bit RCIF in the PIE register is 1, it generates an interrupt request. The bit RCIF is automatically set to 0 when the register RCREG is emptied due to being read by the program.

8.2.4 Communication Speed

The clock that determines the serial port transmission speed is derived from the main microcontroller oscillator. Its value can be adjusted with the SPBRG register and the bit BRGH in the TXSTA register. The transmission velocity in asynchronous mode is

$$v_T = \frac{4^{BRGH}}{64 \times (SPBRG + 1)} \times f_{osc} \, , \qquad (8.2)$$

with v_T being the transmission speed in bits per second, f_{osc} the frequency of the main oscillator in Hz, BRGH can be 0 or 1, and SPBRG is a number between 0 and 255. Table 8.2 shows some calculated transmission speed values following Equation 8.2. The error in the table has been calculated as

$$\text{Error} \, (\%) = \frac{v_{T, \text{calculated}} - v_{T, \text{target}}}{v_{T, \text{target}}} \times 100 \, . \qquad (8.3)$$

If the USART serial port has been programmed in synchronous mode, the transmission speed is given by

TABLE 8.2

Some USART Communication Velocities in Asynchronous Mode and
Percent Error in the Target Velocity

$v_{T\,target}$/(bit/s)	f_{osc}/MHz	BRGH	SPBRG (Decimal)	$v_{T.\,calculated}$/ (bit/s)	Error %
1200	4	0	51	1201.92	0.16
1200	16	0	207	1201.92	0.16
2400	4	0	25	2403.85	0.16
2400	16	0	103	2403.85	0.16
9600	16	0	25	9615.38	0.16
19200	4	1	12	19230.77	0.16
19200	16	1	51	19230.77	0.16
1200	5.0688	0	65	1200.00	0
2400	5.0688	0	32	2400.00	0
9600	5.0688	1	32	9600.00	0

$$v_T = \frac{f_{osc}}{4 \times (SPBRG + 1)}. \tag{8.4}$$

In this communication mode the bit BRGH in the TXSTA register is not
used.

Example 8.1

Basic programming of the USART serial port, in asynchronous mode in a
PIC16F873. The frequency of the main oscillator is 4 MHz.

The program has three parts: First, it is necessary to initiate the serial port in
asynchronous mode establishing a predetermined communication speed. The
second part is to write a subroutine to transmit data. The third part is to write a
subroutine to receive data. This program uses polling input/output.

```
;
; INIT_SCI: Subroutine to initiate the USART serial port in
; asynchronous mode at 19200 bits/s
;
INIT_SCI:
    bsf     STATUS, RP0  ; Select bank 1.
    movlw   0Ch          ; Transmission speed 19200 bit/s.
    movwf   SPBRG
    movlw   0CFh         ; Program RC7/RX pin as input
    movwf   TRISC        ; and RC6/TX pin as output.
    movlw   24h          ; Asynchronous mode, 8 data bits,
    movwf   TXSTA        ; enable transmission, BRGH = 1.
    bcf     PIE1, TXIE   ; Transmission interrupt enabled.
    bcf     PIE1, RCIE   ; Reception interrupt enabled.
```

```
    bcf      STATUS, RP0    ; Select bank 0.
    movlw    90h            ; Enable reception, 8 data bits.
    movwf    RCSTA          ; USART ready to transmit and receive data.
    return                  ; Return.
;
; TXDATA: Subroutine to transmit 8 data bits.
; Data to transmit must be in W register.
;
TXDATA:
    btfss    PIR1, TXIF     ; TXIF=1?
    goto     TXDATA         ; No, wait.
    movwf    TXREG          ; Yes, store data in TXREG.
    return                  ; Return.
;
; RCDATA: Subroutine to receive 8 data bits.
; The received data is stored in W register.
;
RCDATA:
    btfss    PIR1, RCIF     ; RCIF=1?
    goto     RCDATA         ; No, wait.
    movf     RCREG, W       ; Yes, read data.
    return                  ; Return with data in W.
```

8.3 The Synchronous Serial Port in PIC Microcontrollers

Medium-end PIC microcontrollers have a serial port for short distance synchronous communication using the clock signal. The two versions of this serial port are called synchronous serial port (SSP) and master synchronous serial port (MSSP). Any of these ports can be configured to work as a serial synchronous SPI (serial peripheral interface) or as an I²C interface.

In the SSP, the I²C interface can only work as a slave, whereas in the MSSP it can be programmed as a master or as a slave. The SPI was initially designed for the Motorola 68HCxx family of microcontrollers. Both, the SPI and the I²C interface have been developed for communication between devices (microcontrollers, external memory, displays, A/D converters, etc.) located at short distances using a small number of connecting lines. Each of these interfaces is described in further detail in the next sections.

8.3.1 SPI

The SSP or MSSP, when programmed as an SPI, can transmit 8-bit data synchronously and simultaneously. The SPI can work as a master or as a slave. The master is the device that generates the clock signal independently of being a transmitter or a receiver. This communication uses three microcontroller pins that share functions with the parallel port C. These pins are: SDO and SDI for data output and input, respectively; and SCK for the clock. The SCK clock pin is an output pin in the master and an input pin

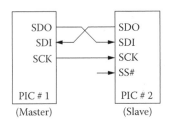

FIGURE 8.18
Two PIC microcontrollers connected using the SPI.

in the slave. A fourth pin called SS# (slave select) can be used for a master device to select one of the several devices programmed as slaves. Figure 8.18 shows the use of these pins in the connection between two PIC microcontrollers.

The SPI uses the special function register SSPBUF to temporarily store either the data to be transmitted or the received data. It also uses the SSPCON and SSPSTAT registers to control the interface. Also, the bits SSPIE and SSPIF in the PIE and PIR registers are used to control and flag the interrupt requests generated by the interface. Figure 8.19 shows the use of the bits in the SSPCON and SSPSTAT registers. These registers are also used to control the I²C interface, but in this case, some of the bits have different meanings.

The bit SSPEN (SSP enable bit) assigns the pins SCK, SDO, SDI, and SS# to the serial synchronous port although they must be defined as inputs or outputs by setting to 1 or 0 the appropriate bits in the TRIS register. Bits SSPM3:SSPM0 in the SSPCON register program the device as a master or as a slave, enable or disable the use of the SS# pin as slave control, and select the clock frequency for the SPI. In the master, the clock signal SCK can be a fraction (1/4, 1/16, or 1/64) of the main oscillator frequency or can be obtained through Timer2. The bit WCOL (write collision detect bit) in the SSPCON register informs when there has been a collision in the SSPBUF register. A collision in this case means an attempt to write data when the previous data was still in the register. The bit SSPOV (receive overflow indicator bit) notifies if data has ceased to be received and stored in SSPBUF. The bit CKP (clock polarity select bit) programs the state of the clock signal (0 or 1) when there are no data to be transmitted.

SSPCON

7	6	5	4	3	2	1	0
WCOL	SSPOV	SSPEN	CKP	SSPM3	SSPM2	SSPM1	SSPM0

SSPSTAT

7	6	5	4	3	2	1	0
SMP	CKE	0	0	0	0	0	BF

FIGURE 8.19
SSPCON and SSPSTAT registers used to control data transmission and reception with the SPI.

The SPI operates in the following way: In the master, the transfer is initiated by writing data in the SSPBUF register. At this point, the data is moved to the shift register SSPSR, as shown in Figure 8.20, and begins to be transmitted immediately by the pin SDO. At the same time, it starts receiving data in the SDI pin. When the reception of data is completed, the bit BF (buffer full status bit) in the SSPSTAT register is set to 1 and the received data is available in the SSPBUF register. Also, the bit SSPIF in the register PIR is set to 1. This allows generating an interrupt request if the SSP interrupt is enabled by having the bit SSPIE in the PIE register set to 1. Once the data is extracted from the SSPBUF register, it is possible to write new data in this register, continuing the process of data transmission and reception.

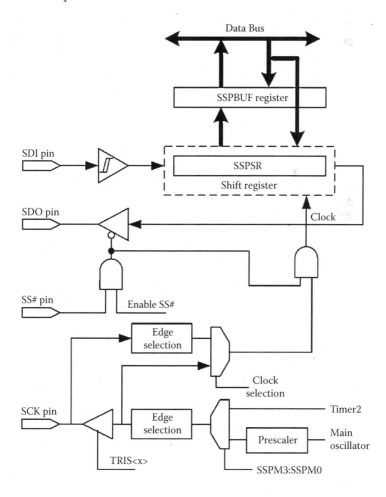

FIGURE 8.20
SPI block diagram.

Figure 8.21 shows the signals for the SPI in master mode. When there is no data to be transmitted, the clock signal SCK is kept in idle state. This idle state can be 0 or 1 depending on how it was programmed in the CKP bit from the SSPCON register. Data bits can be synchronously transmitted with the rising or falling clock edges. This can be programmed with bit CKE (SPI clock edge select bit) in the SSPSTAT register. The sampling of the data signal can be done at half time or at the end of each bit. This is programmed with bit SMP (SPI data input sample phase bit) in the SSPSTAT register.

In a device programmed as a slave, data transmission and reception is initiated when it receives the clock signal. The data to be transmitted must be stored in SSPBUF. When the transmission has finished, the data received is available in the same register, SSPBUF, where it can be read. The reception of the data is flagged with bit BF in the SSPSTAT register being set to 1. But SSPIF in the PIR register is also set to 1. This allows

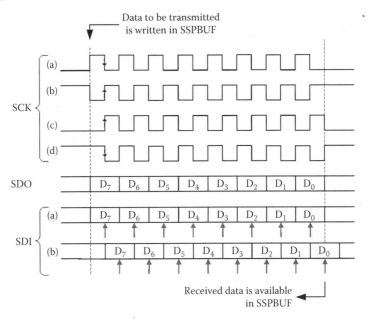

FIGURE 8.21
Signals from the SPI for the device programmed as a master. The transmission starts when the data to be transmitted is written in the SSPBUF register. It is possible to program four variations for the clock signal in SCK: In (a) and (c), the idle state is 0, whereas in (b) and (d) it is 1. It is also possible to program the clock edge to synchronize the data signal. This is indicated by arrows in the signal in SCK. In (a) and (d) data is stable with the falling edge of SCK, whereas in (b) and (c) it happens with the rising edge. The received data signal in SDI can be sampled at times (a) in the middle of each transmitted bit or (b) at the end of each transmitted bit. When the last bit for the input signal is sampled, the received data is available in the SSPBUF register.

generating an interrupt request if the SSP interrupt is enabled by setting the bit SSPIE in the PIE register to 1.

If the slave device is in low-power mode and receives a clock signal, it receives data and transmits the data that was stored in SSPBUF. In this case, the bit SSPIF in the PIR register is set to 1 and if the interrupt is enabled, the PIC services it and wakes up, leaving the low-power mode.

Example 8.2

Data transmission and reception using the SPI port in a PIC16F873 with a clock frequency of 4 MHz.

This example shows how to program the SPI port with the following parameters: programmed I/O technique, master mode, communication clock 1/16 of main oscillator frequency, idle state for SCK is 1, and data transmitted with falling SCK edge. The example also shows the subroutine to receive and transmit data. The program assumes that the data to be transmitted has been previously stored in the DATATX register and the received data is stored in the DATARX register.

```
;
; Serial port SPI programming. Fosc = 4 MHz.
;
        list    p=16f873
        #include <p16f873.inc>
;
DATATX          equ         20h         ; Data to transmit.
DATARX          equ      21h            ; Received data.
;
; INIT_SPI: Subroutine to initiate serial port SPI.
;
INIT_SPI:
        bsf     STATUS, RP0     ; Select bank 1.
        movlw   40h             ; SMP=0, CKE=1, BF=0.
        movwf   SSPSTAT
        movlw   0D7h            ; Program pins RC5/SDO and
                                ; RC3/SCK as outputs
        movwf   TRISC           ; and RC4/SDI as input.
        bcf     STATUS, RP0     ; Select bank 0.
        movlw   31h             ; SSPEN=1, CKP=1, SPI master,
                                ; clock Fosc/16 (250 kbit/s).
        movwf   SSPCON
        bsf     STATUS, RP0     ; Select bank 1.
        bcf     PIE1, SSPIE     ; Disable SPI interrupt.
                                ; SPI ready to transmit and
                                ; receive data.
        bcf     STATUS, RP0     ; Select bank 0.
        movf    DATATX, W       ; Transmit
        movwf   SSPBUF          ; first data.
        return
;
; TRANSFER: Routine to transmit and receive 8-bit data.
; Inputs: Data to be transmitted stored in DATATX.
; Outputs: Received data stored in DATARX.
;
TRANSFER:
```

```
          bsf      STATUS, RP0          ; Select bank 1.
WAITING:
          btfss    SSPSTAT, BF          ; Receive data?
          goto     WAITING              ; No - wait.
          bcf      STATUS, RP0          ; Yes - select bank 0 and
          movf     SSPBUF, W            ; read received data, store it
                                        ; in W and
          movwf    DATARX               ; store it finally in DATARX.
          movf     DATATX, W            ; Data to be transmitted is
                                        ; stored in
          movwf SSPBUF                  ; SSPBUF, initiating the
                                        ; transmission
                                        ; and reception of new data.
          return
          end
```

8.3.2 I²C Interface

The SSP can work as an I²C interface in slave mode. To implement a master I²C interface, the microcontroller needs an MSSP. In the MSSP, the I²C interface can be configured as a master or as a slave. The interface uses pins SDA and SCL for data and clock, respectively. These pins normally share function with port C pins. Figure 8.22 shows the block diagram for the I²C interface. A critical element in this interface is the shift register SSPSR. Each transmitted or received byte—either data, address, or part of an address (when using 10-bit addresses)—must travel through this register. The register SSPSR cannot be seen by the programmer. The communication between this register and the programmer is done by using the special function register SSPBUF. Any data or address to transmit to the bus must be stored in SSPBUF. Any data received by the bus can be read in SSPBUF. Each time that a byte is received or transmitted, the bit SSPIF in the PIR register is set to 1. This generates an interrupt request if bit SSPIE in the PIE register is set to 1.

This interface also uses the SSPADD register that works differently in master or slave modes. In slave mode, the special function register SSPADD stores the address of the device. When the master calls by storing the slave address in the bus, it compares the address sent by the master with the address of the slave stored in SSPADD. If the addresses are the same, the slave carries out an action to respond to the master and establish the communication. This action can simply be to generate an interrupt request (the bit SSPIF in the PIR register is set to 1).

When working as a master, the SSPADD register is used to set the frequency of the clock generated by the master. In this case, the address that the master must store in the bus to call a slave, reaches the shift register SSPSR through the special function register SSPBUF. The master device also has circuits to generate the start and stop conditions. A slave device only has the circuits to detect these conditions. The control of these circuits is done by the special function registers SSPSTAT and SSPCON2.

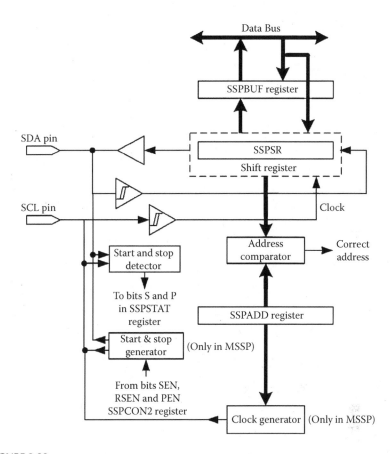

FIGURE 8.22
Block diagram of the SSP interface in I²C mode. Data and addresses are transferred from or to the bus using the SSPBUF register. The shift register SSPSR is invisible to the programmer. The register SSPADD stores the address of the slave device or the communication speed in master mode. The generators for the start and stop conditions, as well as the clock generator, only exist in master devices.

In a master device, the frequency of the clock signal in pin SCL can be obtained as

$$f_{SCL} = \frac{f_{OSC}}{4 \times (SSPADD + 1)}, \tag{8.5}$$

where f_{SCL} is the frequency of the clock in the SCL line of the bus, f_{OSC} is the frequency of the main oscillator in the microcontroller, and SSPADD is the number made of the 7 least significant bits in the SSPADD register (SSPADD<6:0>).

SSPCON

7	6	5	4	3	2	1	0
WCOL	SSPOV	SSPEN	CKP	SSPM3	SSPM2	SSPM1	SSPM0

SSPSTAT

7	6	5	4	3	2	1	0
SMP	CKE	D/A#	P	S	R/W#	UA	BF

SSPCON2

7	6	5	4	3	2	1	0
GCEN	ACKSTAT	ACKDT	ACKEN	RCEN	PEN	RSEN	SEN

FIGURE 8.23
SSPCON, SSPSTAT, and SSPCON2 registers used for data transmission and reception by the I²C interface. The SSPCON2 only exists in devices with an MSSP.

Figure 8.23 shows the bits in the SSPCON, SSPSTAT, and SSPCON2 registers as used by the I²C interface. The SSPCON2 register only exists in those devices that can be masters, that is, they have an MSSP.

Bits SSPM3:SSPM0 in the SSPCON register are used to program the device as a master or slave I²C interface using a 7-bit or 10-bit address. The bit SSPEN (SSP enable bit) assigns pins SDA and SCL to the I²C interface or to the parallel port C. Bit WCOL (write collision detect bit) in the SSPCON register informs about collisions in the SSPBUF register. A collision occurs when the device tries to write data in the SSPBUF when the previous data is still in that register. The bit SSPOV (receiver overflow indicator bit) indicates if the device has ceased to read data received and stored in the SSPBUF register. The bit CKP (clock polarity select bit) is used to control the clock in the slave.

In the SSPSTAT register, bits S (start bit) and P (stop bit) indicate the detection of the start and stop conditions. R/W# (read/write bit information) is a bit that accompanies the address and indicates if the slave is a receiver or a transmitter. UA (update address) is only used in 10-bit addresses indicating that the slave address in the register SSPADD must be updated. The bit BF (buffer full status bit) indicates if the SSPBUFF register is full or empty. D/A# (data/address bit) indicates if the last transmitted or received byte is data or an address. The bit SMP (sample bit) controls the slew rate of signals in high-speed mode (400 kHz).

The SSPCON2 register is used to control the I²C interface when it is configured as a master. Bits SEN (start condition enable bit), RSEN (repeated start condition enable bit), and PSEN (stop condition enable bit) generate the conditions for start, repeated start, and stop. RCEN (receive enable bit) enables the slave reception. ACKEN (acknowledge sequence enable bit) enables the generation of an acknowledgment bit

(A). The value of this bit must be in ACKDT (acknowledge data bit). ACKSTAT (acknowledge status bit) informs of the received acknowledge bit. GCEN (general call enable bit) enables interrupt request for a general call.

9

Analog Input and Output
Signal Acquisition and Distribution

The previous chapters have described how microcontrollers can acquire, process, and generate digital signals that are used, for example, to communicate with other circuits and subsystems. This chapter is focused on the acquisition and generation of analog signals using external modules, peripheral devices, or the devices already integrated inside the microcontroller. We put special emphasis on the basic criteria used for the design of the external modules.

Analog signals are common in measurement and control systems as well as in communications involving human intervention, such as microphones, speakers, and cameras. Analog signals carry information in their amplitude or in their frequency or period, and it is necessary to digitize them first to process them with digital circuits. This requires the use of circuits to adapt their characteristics to the specifications of the digitizing devices. Furthermore, because it is common for analog systems to process signals located farther away from the microcontroller, it is necessary to use the appropriate devices to prevent damage from excessive voltages or currents.

9.1 Structure of a System for Signal Acquisition and Distribution

9.1.1 Basic Functions of Measurement and Control Systems

Most of the signals that need to be measured in a control system are non-electrical signals. Also, the majority of the signals to control, such as the temperature in a room and the position of the head in a printer, as well as the signals to communicate the results to the user, are also nonelectrical. The basic functions for a measurement system are:

- To detect the quantity with a *sensor*
- To process the information
- To communicate the information to the user or another machine

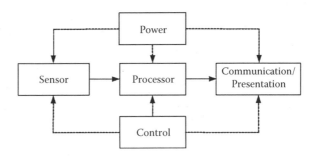

FIGURE 9.1
Basic functions in a measurement system. Changes in the supply voltage and control signals, can affect the output signals for each block.

Two additional functions involved in this process are to supply electrical energy to the system and to control all those functions (Figure 9.1). The processor is normally based on a digital device, for example, a microcontroller that can also control and coordinate those functions. If the goal of the measurement system is to control a variable, after the measurement and processing of the signals, it is necessary to use an *actuator* to convert the electrical output signal into the desired physical action, such as start a heater or turn on a motor.

Typical sensors yield low-amplitude analog signals. Only a few sensors, such as position encoders, offer a digital output that if it has the appropriate voltage levels could be directly connected to the port of a microcontroller. The rest of the sensors need to have their output signals amplified and then digitized by means of an *analog-to-digital* (A/D) *converter*. To adapt the analog signal to the range of expected amplitude at the input of the A/D converter, a *signal conditioner* is used. In some cases, the signal needs additional analog processing, for example, to obtain its rms value or to be demodulated before being digitized. The output of the A/D converter is then processed to obtain the required information, for example, the mean value of a signal during a specific length of time. Often, signals from different sensors are combined to make a decision, for example, for intrusion detection in a given location. The main blocks shown in Figure 9.1 can be expanded as shown in Figure 9.2.

Therefore, there are two main types of functions: conversion and conditioning. Conversion functions create an electrical signal from a nonelectrical signal, or they create a digital signal from an analog signal. Conditioning means to adapt the output signal to the needs of the input for the next stage.

Analog-to-digital conversion can be seen as comparing an unknown voltage (v_x) with a reference voltage (V_{ref}) as shown in Figure 9.3. Direct A/D conversion compares the voltage v_x with fractions of the voltage V_{ref}, each one with a value equal to $L \times V_{ref}/2^N$, with L and N being integer numbers. The comparison can be done simultaneously with all the values

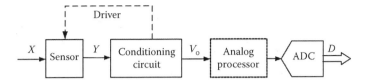

FIGURE 9.2
Front-end in a signal acquisition system: the signal conditioner adapts the input signal coming from the sensor to the input range for the A/D converter. The analog processor can be used to obtain a parameter of interest before digitizing the signal, for example, its rms value, or it can demodulate the signal as needed.

between 0 V and V_{ref} V (*flash converters*), or in consecutive steps using fractions that optimize the decision process (*successive approximation converters*). Successive approximation converters are the type of A/D converters normally found in microcontrollers. The transfer characteristic for an A/D converter with $N = 3$ is shown in Figure 9.3.

Indirect A/D conversion needs an additional circuit that generates a time interval proportional to the input voltage. The length of this time interval is then compared with the time interval generated by the reference voltage using the same circuit. The digital counter measures both time intervals to create the digital conversion. Other indirect converters generate a signal with a frequency proportional to the input voltage, and measure it with a digital counter.

Sensors that embed the information in the frequency, period, time interval, pulse width, duty cycle, phase, and so forth of their output signal, are called *quasidigital sensors*. Although their output is not digital, a simple counter yields a digital code. That is, the sensor itself carries out part of the indirect A/D conversion. Because the result of the counting is an

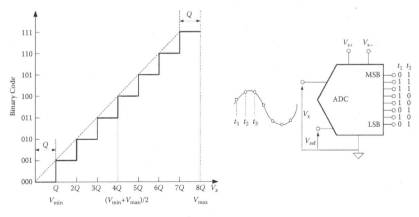

FIGURE 9.3
Direct analog-to-digital conversion process and its transfer characteristic.

integer number, the transfer characteristic for the indirect conversion is also that in Figure 9.3, which can be described by

$$D_x = \text{int}\left(\frac{v_x}{V_{ref}} 2^N\right) + 1,$$

(9.1)

with D_x being the order number for the output code (between 1 and 2^N); int(a) the highest integer equal or less than a, $v_x < V_{ref}$; and N the number of bits in the converter. In an unsigned binary code, the first code is 000...0 and the last code is 111...1. When the value of V_{ref} can be chosen, as happens in several microcontrollers, the accepted range for input voltages depends on the selection of V_{ref}.

The staircase transfer characteristic of an A/D converter causes all input voltages within the range $[V_T, V_T + Q]$ to be assigned to the same code. Here, V_T is any of the threshold voltages after which the assigned code is different, and Q is the so-called *quantization interval*:

$$Q = \frac{V_{ref}}{2^N}.$$

(9.2)

In an A/D converter, Q = 1 LSB. Quantization implies that given a certain output code, it is not possible to know exactly the input voltage that produced that code because several voltages produce the same code. Quantization determines the *resolution* of the system, that is, the lowest change in voltage that will produce a code change.

9.1.2   Dynamic Range

The circuitry between the sensor and the A/D converter is called *analog front-end* (AFE). This AFE needs to be designed taking into account the *dynamic range* (DR) of the measurement, defined as

$$DR_x = \frac{\text{Measurement Range}}{\text{Resolution}} = \frac{x_{max} - x_{min}}{\Delta x}.$$

(9.3)

It is possible to define the dynamic range for the input or output in any stage as

$$DR = \frac{V_{max} - V_{min}}{\Delta V}$$

(9.4)

with ΔV being the resolution in that particular stage.

The resolution at the input of a stage is the lowest change that produces a change in the output. The resolution at the output of a stage is the resolution at the input multiplied by the voltage gain for the stage. Therefore, stages with unity gain have the same resolution at their input and at their output. In an A/D converter,

$$DR_{A/D} = \frac{V_{max} - V_{min}}{Q} = \frac{(2^N - 1)Q}{Q} \approx 2^N. \tag{9.5}$$

The design of the analog front-end is based on the dynamic range of the measurement because this value determines the number of bits for the A/D converter. Because the A/D has specific requirements regarding its range and voltage values, the front-end stage must adapt the sensor output to these specific requirements. Therefore, in the design of the stages, shown in Figure 9.2, choosing the sensor and the A/D converter determines the intermediate stages. In a correct design, the dynamic range at the input of each stage must be equal or higher than the dynamic range of the preceding stage, as shown in Figure 9.4. *Noise* is the rms value of the signal measured at the output of a stage when its input is zero. Noise determines the ultimate resolution of the measurement system. If a stage has a very large gain, the noise of the next stage is insignificant compared with the noise produced by that previous stage.

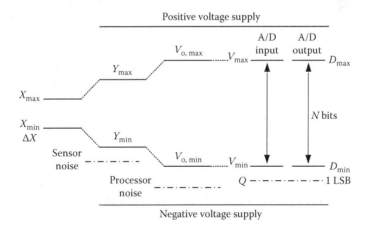

FIGURE 9.4
The A/D converter must have a bigger dynamic range than the sensor. Also, the stages between them must have a bigger dynamic range than the A/D converter.

Example 9.1

We wish to measure a temperature between −40°C and 60°C with a resolution of 0.5°C. How many bits does the A/D converter need? If we use a sensor with a sensitivity of 1 mV/°C and an A/D converter with an input range between 0 V and 5 V, how much gain is necessary? Calculate the effective resolution obtained when measuring temperature without the use of amplification.

The measurement range is

$$DR = \frac{60°C - (-40°C)}{0.5°C} = 200$$

Using Equation 9.5,

$$N = \frac{\log 200}{\log 2} = 7.6 \text{ bit}$$

Therefore, the A/D converter should have 8 bits.

The sensor voltages at the limits of the measurement range are −40 mV and 60 mV. To condition these voltages to the 0 V to 5 V range at the input of the A/D converter, a gain of it is necessary

$$G = \frac{5 \text{ V} - 0 \text{ V}}{60 \text{ mV} - (-40 \text{ mV})} = 50 .$$

The analog processor also needs to shift the voltage by 40 mV in order for the −40°C to correspond to 0 V, which is the minimum voltage at the input of the A/D converter. This level shift must be introduced before the signal is amplified.

Without amplification, given that the resolution of the A/D converter is $5 \text{ V}/2^8$ = 19.5 mV and the sensitivity of the sensor is 1 mV/°C, the effective resolution is 19.5°C. Also, if the output voltage is not shifted, the negative voltages that correspond to the temperatures between −40°C and 0°C will not be digitized.

9.1.3 Bandwidth

In addition to conditioning the amplitude and range, the front-end must adapt the frequency of the signal to be digitized to the specifications of the A/D converter. Each block in the analog front-end must have a frequency response adequate to the signal to be processed. The *bandwidth* of a signal is defined as the difference between the maximum and minimum frequencies that bring significant information to the signal. The *central frequency* is defined as the geometric mean between the extreme frequencies:

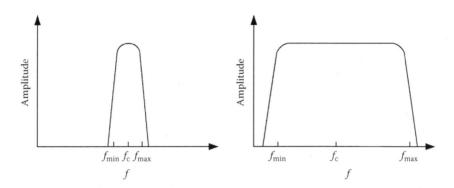

FIGURE 9.5
Narrowband signal (left) and wideband signal (right). The criteria to determine f_{max} and f_{min} depend on each specific signal. The maximal amplitude does not necessarily correspond to f_c. The amplitude is not necessarily uniform between f_{max} and f_{min}.

$$f_c = \sqrt{f_{max}f_{min}} \qquad (9.6)$$

Depending on the relationship between these frequencies it is possible to differentiate between *narrowband signals* and *wideband signals*. Narrowband signals are those signals in which $f_c > f_{max} - f_{min}$, whereas wideband signals are those signals in which $f_c < f_{max} - f_{min}$. Figure 9.5 shows examples of these two types of signals.

From this classification it is possible to see how, for example, a signal between 1 MHz and 2 MHz is a narrowband signal and a signal between 1 Hz and 100 Hz is a wideband signal. Wideband signals are more difficult to process because the processor specifications must be kept uniform within the bandwidth. Signals that only have very low frequency components, less than 0.1 Hz, are called *dc (direct current) signals*, whereas those that have higher frequencies are called *ac (alternate current) signals*.

The *circuit bandwidth* is the frequency region in which the response to an input signal is within ±3 dB of the frequency response in the middle of the frequency interval as shown in Figure 9.6. The *central frequency* is defined as the geometric mean of the frequencies at −3 dB.

When stages or circuits are connected without any intervening series capacitors, the connection is called *dc coupling*. When a series capacitor is used, the connection is called *ac coupling*. In this case it is necessary to take into account that series capacitor will attenuate low-frequency signals.

9.1.4 Signal Sampling

In direct A/D converters, the digitized value is equal to the value that the input voltage has during the conversion process. For this reason,

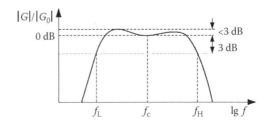

FIGURE 9.6
Circuit bandwidth is defined with respect to the frequencies with a gain within ±3 dB compared to the response at the central frequencies.

the voltage must be kept constant while it is being digitized. This can be achieved by using a *sample-and-hold amplifier* (SHA). The number of samples that need to be taken in a unit of time depends on the type of processing that will be done. The best choice would be to use an ideal digital interpolating filter. In this case, the sampling frequency is given by the *Nyquist criterion*, which specifies that the sampling frequency has to be higher than twice the signal bandwidth. If the sampling frequency is less that the Nyquist frequency, *alias* signals will appear, as shown in Figure 9.7.

For periodic signals it is possible to take samples at different periods as shown in Figure 9.8. This reduces the sampling rate required. This technique is called *repetitive sampling* and is widely used in digital oscilloscopes.

9.1.5 Architectures for Signal Acquisition: High-Level and Low-Level Mutiplexing

To digitize several signals, if the A/D converter is fast enough, we do not need to use one A/D converter for each signal. Instead, a single A/D converter can be shared by all the signals using an *analog multiplexer*. This is a set of analog switches with a common output. At any given time, only one of the switches is closed—the one whose input will be digitized.

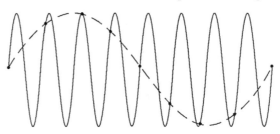

FIGURE 9.7
Aliasing appearing when sampling at a frequency below the Nyquist frequency.

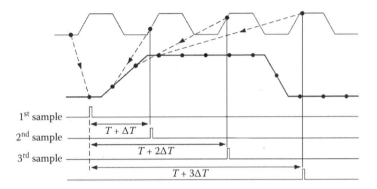

FIGURE 9.8
Repetitive sequential sampling: the signal is sampled at increased times within successive cycles for the periodic input signal.

The location of the multiplexer in the measurement system leads to several architectures for data acquisition. The most common architectures use high-level or low-level multiplexing. Systems with *high-level multiplexing*, like that shown in Figure 9.9, only accept large signals and have a programmable amplifier with relatively low gain (1, 2, 4, 8) after the multiplexer. Therefore, each signal must be conditioned before being multiplexed, thus increasing the cost of the design. The advantage of these systems is that their switching speed, and therefore the sampling rate, can be fast because each signal has been individually

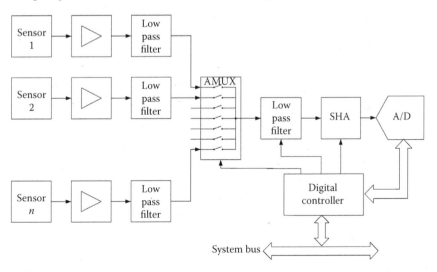

FIGURE 9.9
High-level multiplexing: each signal must be conditioned before being multiplexed. SHA, sample-and-hold amplifier.

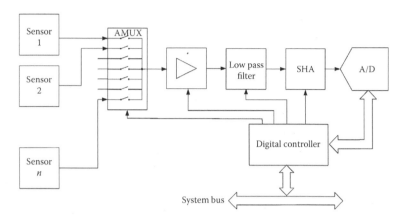

FIGURE 9.10
Low-level multiplexing: signals do not have to be conditioned before being multiplexed, but the switching speed between channels is slower because it is necessary to wait for the amplifier and filter outputs to settle.

adapted to the specifications of the multiplexer. Those microcontrollers that include an A/D converter with several analog inputs (normally up to 16 inputs) have the multiplexer integrated in the chip but not the amplifier or the filter that must be connected externally. Some multi-channel A/D converters do not incorporate the amplifier or the filter either. If it is required to sample several signals simultaneously, it is necessary to have a sample-and-hold amplifier for each one of them before they are multiplexed.

Systems with *low-level mutiplexing*, like that shown in Figure 9.10, include a programmable gain amplifier with gains between 1 and 500. The input signals can be smaller than those required by high-level systems, but the errors from the multiplexer will be added to each signal and will be amplified. Also, when switching from one channel to the next, it is necessary to wait until the output from the amplifier and the filter are settled. This reduces the maximum speed at which the channels can be scanned. These low-level acquisition systems are available as peripheral integrated circuits.

9.2 The Front-End in Data Acquisition Systems

The front-end in data acquisition systems must adapt the voltage, current, and power levels of the input signal to the A/D converter specifications, while maintaining the signal bandwidth.

9.2.1 Attenuators

When the voltage amplitude for
the input signal exceeds the maximal voltage allowed by the A/D
converter, the input signal needs
to be attenuated. Figure 9.11 shows
the general structure of a voltage
attenuator made of two impedances, Z_1 and Z_2. The figure also
assumes that the equivalent output impedance for the source is Z_o
and the equivalent input impedance is Z_i. In these conditions, the
attenuation is

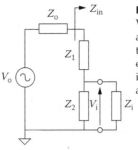

FIGURE 9.11
Voltage attenu-
ator connected
to a stage whose
equivalent
input imped-
ance is Z_i.

$$A = \frac{V_i}{V_o} = \frac{Z_2 \| Z_i}{Z_o + Z_1 + Z_2 \| Z_i} \approx \frac{Z_2}{Z_1 + Z_2},$$ (9.7)

with the symbol $\|$ representing the parallel combination of two imped-
ances. This approximation is valid as long as $Z_{in} \gg Z_o$ and $Z_i \gg Z_2$. Z_{in}
represents the equivalent input impedance for the attenuator once it is
connected to the next stage. That is, Z_{in} is the equivalent impedance seen
by the signal source. Its value is

$$Z_{in} = Z_1 + Z_2 \| Z_i.$$ (9.8)

Accomplishing the criterion "much higher than" depends on the A/D
converter resolution. The criterion will be fulfilled if the effect does not
influence the A/D converter. This means that the effect of having a finite
impedance value is less than Q. The evaluation of a parameter that is ideally zero is done in a similar way.

A dc voltage can be attenuated with just two resistors connected, as
shown in Figure 9.12a, resulting in an attenuation equal to

$$A = \frac{V_i}{V_o} = \frac{R_2 \| R_i}{R_o + R_1 + R_2 \| R_i} \approx \frac{R_2}{R_1 + R_2},$$ (9.9)

whereas the equivalent input resistance for the attenuator is

$$R_{in} = R_1 + R_2 \| R_i.$$ (9.10)

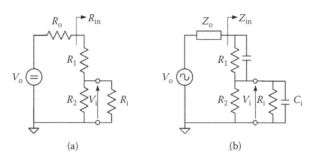

FIGURE 9.12
Voltage attenuator for (a) dc signals and (b) ac signals.

These two equations are used to calculate the values of R_1 and R_2. The tolerance in their values and the effect of the temperature coefficient make the actual value of A unknown. To find out this value it is necessary to calibrate the system (Section 9.4).

Example 9.2

Design a resistive attenuator with an input resistance of 1 MΩ that will allow measuring a voltage of 42 V with a circuit that admits 5 V having an input resistance of 1 MΩ.

The attenuation condition (Equation 9.9) and input resistance (Equation 9.10) give

$$A = \frac{5 \text{ V}}{42 \text{ V}} = \frac{R_{eq}}{R_1 + R_{eq}}$$

$$R_{in} = R_1 + R_{eq} = 1 \text{ M}\Omega$$

with $R_{eq} = R_1 || R_2$. From these equations, R_1 = 881 kΩ and R_2 = 135 kΩ. When using resistors with 1% tolerance, the nominal values are R_1 = 887 kΩ and R_2 = 133 kΩ. These values slightly increase the attenuation but this assures that the signal range will not be exceeded.

If the same circuit was used to attenuate ac signals, the equivalent input capacitance (in parallel with R_i) would attenuate more higher frequency signals than low frequency signals. This in turn prevents one from determining the value of the input voltage for wideband signals. This problem can be prevented by using the circuit shown in Figure 9.12b that introduces a capacitor C_1 in parallel with R_1. Choosing C_1 so $R_1C_1 = (R_2||R_i)C_i$, the attenuation becomes

$$A = \frac{V_i}{V_o} = \frac{Z_{eq}}{Z_o + Z_1 + Z_{eq}} \approx \frac{Z_{eq}}{Z_1 + Z_{eq}} = \frac{R_2 \| R_i}{R_1 + R_2 \| R_i}, \tag{9.11}$$

$$Z_{in} = \left(R_1 + R_2 \| R_i \right) \| \left(C_1 \oplus C_i \right). \tag{9.12}$$

with Z_{eq} being the parallel combination of Z_2, R_i, and C_i, and the symbol $\oplus$ representing the serial combination of two capacitors. From Equation 9.11 it can be seen that the attenuation is constant. This attenuator is then called a *compensated attenuator*. These two equations and the condition for compensation can be used to calculate the values of R_1, R_2, and C_1.

Example 9.3

Design an attenuator to divide a voltage signal by a factor of 10 at any frequency, with an input resistance of 10 MΩ at low frequency when connected to a data acquisition system that has an input impedance of 1 MΩ || 125 pF.

Using Equation 9.11 to obtain the necessary attenuation yields:

$$A = \frac{R_{eq}}{R_1 + R_{eq}} = 0.1$$

From here, $R_1 = 9R_{eq}$. The condition for the time constants implies $C_1 = C_i/9 = 125$ pF/9 ≈ 14 pF. To obtain the desired input impedance at low frequency:

$$R_{in} = R_1 + R_{eq} = 10 \text{ M}\Omega = 10 \, R_{eq}$$

Therefore, $R_{eq} = 1$ MΩ implies $R_2 = \infty$ (open circuit). With this, $R_1 = 9$ MΩ. Standard values are $R_1 = 8.98$ MΩ (±0.5%) or $R_1 = 9.09$ MΩ (±1%).

If the output impedance of the source (Z_o) in Figure 9.12 is not low enough, for example, because the input impedance Z_{in} is low at the signal frequency due to parasitic capacitances, the previous approximations are no longer valid. There is a so-called *voltage loading effect*, and the attenuation increases as the frequency increases. The general situation is shown in Figure 9.13a. For ac voltages coming from a resistive source, it becomes the circuit shown in Figure 9.13b, which can be further simplified to the circuit in Figure 9.13c with $R_{eq} = R_o || R_{in}$. The voltage at the input is not V_o but V_{in}, as shown below:

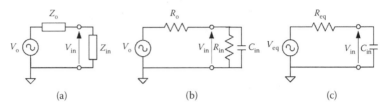

FIGURE 9.13 Equivalent circuits used to analyze voltage-loading effects for ac signals.

$$V_{in} = V_o \frac{Z_{in}}{Z_o + Z_{in}} = V_o \frac{R_{in}}{R_o + R_{in}} \frac{1}{1 + j2\pi f \left(R_o \| R_{in} \right) C_{in}} = V_{eq} \frac{1}{1 + j2\pi f R_{eq} C_{in}}. \quad (9.13)$$

The difference between the measured and applied voltages is the absolute error. When it is divided by the input voltage, it becomes the relative error. If this error needs to be lower than a predetermined value (ε), it is then necessary to meet

$$\frac{\left\| V_{in} \right| - \left| V_o \right\|}{\left| V_o \right|} < \varepsilon, \quad (9.14)$$

giving the condition

$$2\pi f R_{eq} C_{in} < \frac{\sqrt{A_0^2 - 1 - \varepsilon^2 + 2\varepsilon}}{1 - \varepsilon} \approx \frac{\sqrt{A_0^2 - 1 + 2\varepsilon}}{1 - \varepsilon} = \frac{\sqrt{2\varepsilon}}{1 - \varepsilon}. \quad (9.15)$$

where $A_0 = R_{in}/(R_o + R_{in})$ is the dc attenuation. The first approximation is acceptable when $\varepsilon \ll 1$ and the last step assumes A0 very close to 1. Equation 9.15 allows us to determine, for example, the maximal frequency for the input voltage that yields a relative error smaller than ε.

Example 9.4

Calculate the maximal frequency of a sine signal generated by a source with an internal resistance of 600 Ω, when it is connected to a circuit with an input impedance of 10 MΩ||100 pF for the loading effect to be negligible in a system with 12 bits of resolution.

For the loading effect to be negligible, the undesired attenuation experienced by the signal must be less than 1 LSB. 1 LSB = $V_{FS}/2^{12}$. The difference between the source voltage and the voltage at the input of the circuit will be maximum when the voltage is maximum. Therefore, $\varepsilon = 1/2^{12}$. Condition 9.15 can be rewritten as

$$f < \frac{1}{2\pi R_{eq} C_{in}} \frac{\sqrt{2\varepsilon}}{1-\varepsilon}.$$

With the component values in this example, $R_{eq} \approx 600~\Omega$ and $C_{in} = 100$, we can find $f < 58.6$ kHz.

9.2.2 Amplifiers

Amplifiers are used to adapt the dynamic range, levels, and terminal configuration between stages, as well as to offer a high input impedance to avoid voltage-loading effects. The configuration of the terminals of a stage refers to the relationship between the input terminals and the reference voltage in that stage. The reference voltage is called common terminal, 0 V, signal ground, or electric ground (Figure 9.14). When one of the two terminals used to measure a voltage signal is directly connected to the common terminal, we have a *single-ended voltage*. When none of the two terminals used to measure a voltage signal is directly connected to the common terminal, we have a *differential signal* if the following relationship holds true:

$$\left. \begin{array}{l} v_H = v_c + \dfrac{v_d}{2} \\[2mm] v_L = v_c - \dfrac{v_d}{2} \end{array} \right\}. \tag{9.16}$$

v_c is called the *common-mode voltage* and v_d is called the *differential-mode voltage*. For example, in a system powered between 0 V and 5 V, if a

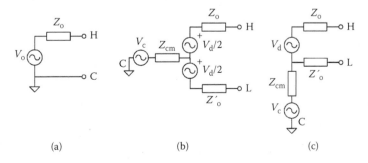

(a)　　　　　　(b)　　　　　　(c)

FIGURE 9.14
Classification of signals depending on the configuration for their terminals: (a) single-ended, (b) differential, and (c) pseudodifferential. Z_{cm} is the common-mode impedance. This value can be zero.

differential voltage has a common-mode voltage of 2.5 V, it is possible to process positive and negative differential voltages because only their sign changes without exceeding the supply voltages.

When condition 9.16 is not met, and therefore the signal is not symmetrical, the signal is said to be a *pseudodifferential voltage*. For differential and pseudodifferential voltages, if $Z_o = Z_o'$, the signal is said to be *balanced*. When Z_{cm} is very high, the signal is said to be a *floating* or *off-the-ground signal*. Both differential and pseudodifferential signals can be described by their three terminals: high terminal, low terminal, and common terminal. A single-ended signal is described by two terminals only: high and common.

A linear amplifier outputs a voltage proportional to the voltage difference between its high-input terminal and low-input terminal. Similar to signal configurations, the input of the amplifiers can be single-ended, differential, or pseudodifferential, as shown in Figure 9.15. In a *single-ended input*, the low terminal is the common terminal (0 V) of the power supply for the amplifier. This is normally connected to the chassis of the equipment (chassis ground), and in turn to the protective earth conductor for the power distribution system. In a *differential input*, none of the two input terminals (high, low) is connected to ground, and the input impedance between each terminal and the ground is about the same. In a *pseudodifferential input*, the low-input terminal is connected to the ground of the amplifier, but the ground of the amplifier is not directly connected to the chassis ground nor to the protective earth conductor. If both Z_L and Z_H are very high, the input is said to be *floating*.

To minimize voltage-loading effects, the impedance between the two measuring terminals must be high enough. To avoid the influence of the common-mode voltage on the output voltage in a differential amplifier, Z_C must be as high as possible and equal for each one of the input terminals. Otherwise, Z_o and Z_C on one side, and Z_o' and Z_C' on the other side

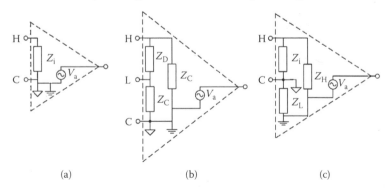

(a)   (b)   (c)

FIGURE 9.15

Amplifiers classified according to their input: (a) single-ended, (b) differential, and (c) pseudodifferential. If Z_L and Z_H are very high, the input is floating.

constitute voltage dividers with different attenuation values, and a common-mode voltage from the signal source produces a differential voltage at the input of the amplifier. In pseudodifferential amplifiers, Z_L and Z_H must be as high as possible. Obviously, a differential or pseudodifferential signal cannot be connected to a single-ended input, as this one does not have enough input terminals. Furthermore, when connecting a single-ended signal to a single-ended input, it is necessary to connect it with the correct polarities. This is less important in differential or pseudodifferential inputs. In any case, it is critical to ensure that common-mode voltages will not significantly contribute to the voltage being amplified.

When the input terminals of an amplifier are connected to ground, its output voltage is not zero as would be expected. Instead, the output is a dc signal whose value depends on the amplifier gain and the resistances between the inputs and ground. To analyze these effects, the amplifier can be modeled by adding a voltage source called *offset voltage* and a dc current source between each input terminal and ground. Figure 9.16 shows the equivalent circuit when connecting a differential voltage to an *instrumentation amplifier*. The instrumentation amplifier is a differential amplifier with very high input impedances. The presence of the dc current sources requires a low resistance path between each input and the ground. For this reason, the common terminal for the signal is normally connected to the amplifier ground. Furthermore, it is not possible to connect the signal to the amplifier using just a capacitor in series with each terminal, because the capacitors would be charging and would end up saturating the amplifier.

When the input currents are low enough and the voltage supply is well filtered, the voltage at the output of the amplifier can be approximated as

$$v_a = G\left(1 + \varepsilon_G\right)\left(v_d + \frac{v_c}{\mathrm{CMRR}} + V_{io}\right) + e_{\mathrm{NLG}} ,\qquad (9.17)$$

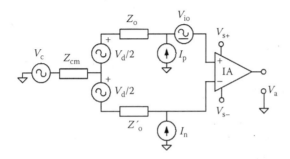

FIGURE 9.16
Equivalent circuit when connecting a differential signal to an instrumentation amplifier. The signal and the amplifier are connected to the same ground.

with ε_G being the relative gain error, e_{NLG} the error for nonlinear gain, and CMRR the *common-mode rejection ratio*. The CMRR is defined as the ratio between the output voltage produced by a differential input voltage and the output voltage when the same input voltage is applied in common mode. If the common-mode input impedances are not very high or they are unbalanced, the effective CMRR is lower than the CMRR for the amplifier itself. Gain errors and offset voltages normally found in instrumentation amplifiers do not preclude dynamic ranges of 100 dB or higher. However, they make it very difficult to achieve without calibration the required accuracy for systems with more than 10 bits. When calibration is used, the accuracy extends to 14 bits at low frequency or for narrowband signals. Wideband signals are limited to an accuracy of 12 bits. The gain of an amplifier decreases after a certain frequency. This frequency decreases when the gain increases. Figure 9.17 shows this effect for an instrumentation amplifier. For most amplifiers, the reduction in gain with the increase in frequency can be described as

$$G(f) = \dot{G}_0 \frac{f_a}{f_a + jf}, \qquad (9.18)$$

with G_0 being the gain at low frequency and f_a the cutoff frequency at −3 dB. From a practical point of view, this dependence means that given a maximum accepted error equal to ε, the maximum allowed frequency for a sine signal at the input is

$$f_{max} < f_a \frac{\sqrt{2\varepsilon - \varepsilon^2}}{1 - \varepsilon} = f_a \frac{\sqrt{2\varepsilon}}{1 - \varepsilon}. \qquad (9.19)$$

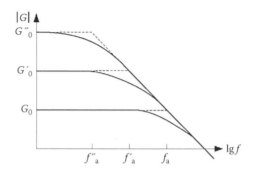

FIGURE 9.17
Relationship between gain and frequency for a typical instrumentation amplifier. The frequency at which the gain starts to decrease is lower for higher values of gain at low frequencies.

Example 9.5

A signal is amplified with an instrumentation amplifier with a gain-bandwidth (GBW) product equal to 1 MHz and a gain of 1000. What is the maximal frequency for the attenuation not to be perceived by a system with 12 bits of resolution?

With $G = 1000$, the –3 dB cutoff frequency is $f_a = 1$ MHz/1000 = 1 kHz. Using Equation 9.19 with $\varepsilon = 1$ LSB/V_{FS},

$$f < f_a \frac{\sqrt{2\varepsilon}}{1-\varepsilon} = 1 \text{ kHz} \sqrt{2 \times 2^{-12}} = 22 \text{ Hz}.$$

This is a very low frequency, but it is necessary to realize that –3 dB is an attenuation of 30% and the maximum attenuation accepted in this example is only 0.024%.

9.2.3 Input Protections and Filters

The maximum voltage that can be directly applied to the input of any electronic device without causing damage is always limited, at least to less than the supplied voltage. The maximum current at the pins of the device is also limited, for example, to less than 20 mA for most of the PIC pins, and less than 10 mA or 1 mA for amplifiers and multiplexers, respectively, depending on their technology. For this reason, it is necessary to add protection circuits for the pins that accept connections to external devices. As shown in Figure 9.18, current limiters are connected in series, and voltage limiters are connected in parallel with the input to protect. For differential inputs it is necessary to add a current limiter in series with the low-input terminal and a common-mode voltage limiter between each input terminal, and ground. When the voltage limiter is engaged, the voltage at the input of the device is constant and independent of the signal. Therefore, the device is not damaged, but information is lost.

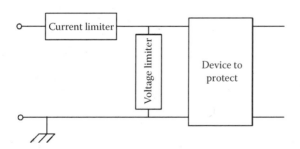

FIGURE 9.18
Current limiters are connected in series with the input to protect, whereas voltage limiters are connected in parallel.

Current can be limited by linear resistors or nonlinear resistors such as positive temperature coefficient (PTC) thermistors. Voltage limiters employ Schottky or Zener diodes or metal oxide varistors (MOV) when the protection threshold is 15 V or higher. The 2Prom™ devices from Tyco use both types of protections. The nominal power for current and voltage limiters is chosen depending on the maximal power that the source can deliver. If it is necessary, the protection can be implemented in two stages: the first for higher power and the second for lower power. Figure 9.19 shows two examples of the first type and one example of the second type. The resistance of the series elements increases the loading effects when measuring voltage. The parasitic capacitance of the elements connected in parallel reduces the bandwidth and the common-mode input impedance in differential inputs.

Analog inputs also need interference filters because the amplitude of these interfering signals from other devices can be high enough to saturate the amplifier's input. These filters must be passive because active filters do not accept voltages beyond the voltage supply rails of their integrated circuits. The series resistor of a first-order low-pass RC filter can be the current limiting resistor. To achieve a high enough capacitance, we can connect a capacitor in parallel with the voltage limiter. The maximum frequency when accepting a relative error ε can be calculated with Equation 9.19, with f_a being the -3 dB filter cutoff frequency ($f_a = 1/(2\pi RC)$).

When the signal to be processed is connected to the ground in a place different than the ground connector for the power supply, the voltage difference between these two points can be higher than the supply voltage. This situation is very common in industrial environments and when measuring in systems with devices far away from one another. In these cases, the voltage limiters would be constantly engaged and information could not be acquired. This problem can be solved by breaking the ohmic

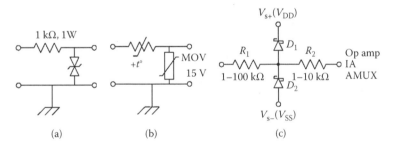

FIGURE 9.19
First-level protection for power values around 1 W. (a) Power resistor and two Zener diodes. (b) A PTC resistor and a varistor (MOV). (c) Second-level protection for extremely vulnerable components: R_1 is not necessary if there is a first-level protection stage; R_2 is only needed if the voltage drop across the Schottky diodes (0.3 V) is still dangerous for the device to protect.

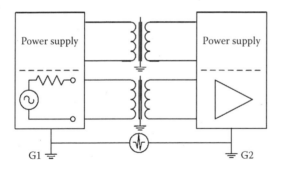

FIGURE 9.20
When the voltage difference between the two ground connections is too high, it is necessary to use isolation methods for the signal and the power supply. Isolation will prevent the excessive voltage difference between grounds to originate a dangerous current in the circuit.

continuity between the signal and the amplifier using a linear isolator or an isolation amplifier for the analog signal. Alternatively, it is also possible to digitize the signal with a system whose voltage common terminal is not connected to ground and transmit the digital signal using optocouplers or other digital isolators. In any case, it is necessary to break the ohmic continuity for both the signal circuit and the power supply circuit, as shown in Figure 9.20.

9.2.4 Analog Multiplexers

An analog multiplexer is a circuit built from a set of analog switches with a common output pin. They are activated in such a way that at any given time, only one of the inputs is connected to the output. Figure 9.21 shows the structure of a multiplexer for single-ended signals and one for differential signals. In multiplexers for differential signals, two switches must close at the same time. The multiplexers found in some microcontrollers allow for measuring the difference between any two signals, not only predetermined pairs of signals, as shown in Figure 9.21b.

The switches of an ideal multiplexer have zero resistance when closed and infinite resistance when open, thus making input and output totally isolated. The switches in real multiplexers are made of CMOS transistors that have a resistance different than zero when closed (R_{ON}) and a finite capacitance (C_{DS}, C_{OFF}, C_{ISO}) between input and output. R_{ON} creates an additional voltage loading effect (additional because it adds to that of the output resistance of the source) called *insertion loss*. This value is sometimes expressed in decibels for a predetermined load resistance that should be the equivalent input resistance for the following stage. Multiplexers that incorporate an additional resistance for overcurrent protection in series with each switch have higher insertion losses.

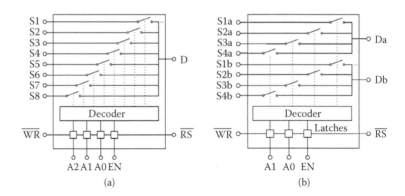

FIGURE 9.21
Functional structure for an analog multiplexer for (a) single-ended signals and (b) differential signals.

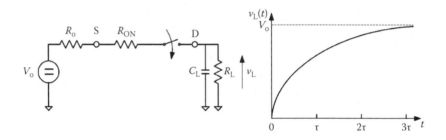

FIGURE 9.22
Equivalent circuit for an analog switch that closes, and evolution of the output voltage. C_L is the equivalent capacitance between the output of the multiplexer and ground. It includes the multiplexer output capacity and the equivalent input capacitance of the following stage.

Furthermore, R_{ON} and R_o introduce a delay when switching between channels because they limit the charging current for the equivalent capacitance connected at the output of the multiplexer. With the equivalent circuit from Figure 9.22, assuming a low-frequency signal at the input of the switch, when the switch is closed, the output voltage is:

$$v_L(t) = V_o\left(1 - e^{-t/\tau}\right), \tag{9.20}$$

with $\tau = [(R_o + R_{ON}) \parallel R_L] C_L$. If the relative difference between the output voltage at a certain time and its final value has to be lower than ε, it is necessary to wait a certain time (t_ε) after closing the switch. This time is given by

$$t_\varepsilon = -\tau \ln \varepsilon. \tag{9.21}$$

This time limits the maximum velocity at which the channels can be scanned. This time is not listed in the multiplexer's data sheet because it depends on external factors, such as R_o and C_L.

Example 9.6

Consider a multiplexer with each channel protected against currents with an equivalent channel resistance of 1 kΩ, and with an output connected to a load of about 100 pF and a very high resistance. Calculate the time necessary to wait after switching one channel for the output to be measured correctly with a 12-bit system.

Using Equation 9.21 with τ = 1 kΩ × 100 pF = 100 ns and ε = 1LSB/2^{12} yields

$$t_\varepsilon = -\tau \ln \varepsilon = \left(-100 \text{ ns}\right) \ln 2^{-12} = 832 \text{ ns}$$

This time is much higher than the switching times between channels in a common multiplexer.

The finite isolation in the multiplexer switches produces *static crosstalk*: when a switch is open, part of the signal at its input is transferred to the output. The output should only have the signal whose switch is closed, and instead it receives contributions from the other signals. As shown in the circuit from Figure 9.23, as the frequency of the signal connected to the open switches increases and the resistance of channel 1 (R_{i1} and R_{ON1}) increases, the crosstalk of channel 2 over channel 1 also increases.

Another limitation from multiplexers is the time elapsed between giving an order to change a channel and when the channel is effectively closed. This is called *switching time*. Also, the signals for controlling channels can produce spurious voltages at the output. When switching channels there is a short time interval in which none of the channels are closed, causing the output to hold the voltage that it had when the previous channel was selected. The *throughput rate* specified by the manufacturer sometimes refers to the ideal situation of having the same continuous voltage applied to each channel and, therefore, the only limiting effect is the output settling time without additional delays or evaluating the effects of crosstalk that occur when switching ac signals.

9.2.5 Anti-Alias Filters

To avoid the creation of alias signals when sampling, it is necessary that the amplitude of any signal at a frequency equal to or higher than the

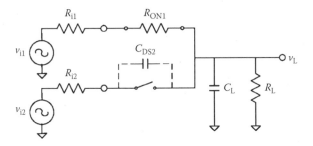

FIGURE 9.23
Equivalent circuit used to analyze static crosstalk in an analog multiplexer. When channel 1 is selected, channel 2 and the rest of the open channels influence the output voltage because the isolation of the open switches is finite.

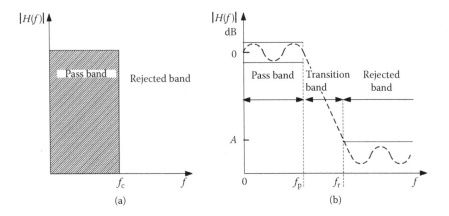

FIGURE 9.24
Frequency response for a low-pass filter: (a) ideal and (b) real.

Nyquist frequency be unnoticed by the A/D converter. The circuits that discriminate signals depending on their frequency are called *filters*. Filters that attenuate high-frequency signals, as is required to avoid alias signals when sampling, are called *low-pass filters*. An ideal filter such as the one shown in Figure 9.24a totally eliminates undesired signals without modifying the frequency or the phase of the signals to pass. In a real filter (Figure 9.24b) the attenuation in the rejected band is limited and the response in the pass band is not constant with the frequency. The transition between these two regions is not abrupt but slow. Furthermore, the phase change that is introduced may not be proportional to the frequency resulting in waveform distortion. An ideal filter would respond immediately to an abrupt change in amplitude without overshooting. A real filter has a frequency response that is very different from the frequency response of an ideal filter.

Filters can be characterized by their order (*n*). The attenuation of common filters increases in 20*n* (dB) as the frequency increases in one decade. The attenuation in the transition band depends on the type of filter selected, which receives a specific name depending on the type of polynomial selected for the denominator of its transfer characteristic. The most common filters are Butterworth, Chebyshev, and Bessel. Butterworth filters offer the flattest response in the pass band. Chebyshev filters offer higher attenuation in the transition band, but show amplitude ripple in the pass band and a nonlinear phase shift. Bessel filters have an almost linear phase shift and do not have overshoot for step inputs, but they have an increased attenuation in the pass band and lower attenuation in the transition band. The designer needs to select the type of filter depending on the specifications given by the application. Once this has been selected, there are several programs to determine the order of the filter as a function of the –3 dB cutoff frequency in the band pass, the sampling frequency, and the desired signal-to-noise ratio. Several different circuits may be used to implement the filter. Filter realization is also supported by software programs such as FilterWizard® (Analog Devices), FilterCAD® (Linear Technologies), Filterlab® (Microchip), or FilterPro® (Texas Instruments).

Example 9.7

Design an anti-alias filter for a signal with a –3 dB bandwidth of 70 Hz and a signal-to-noise ratio of 40 dB when it is sampled at 1500 Hz with a 12-bit converter.

The cutoff frequency of the filter is chosen to be 70 Hz (in order not to modify the signal). Using the "Anti-Aliasing Wizard" tool from Filterlab®, introducing the sampling frequency, the number of bits, and the input signal-to-noise ratio, the result is a second-order filter, type Butterworth with an attenuation of 41.3 dB at 750 Hz (1500 Hz/2).

As an alternative method, for the noise amplitude to be lower than 1 LSB at half of the sampling frequency in a 12-bit system, it is necessary that the noise is approximately 72 dB below the signal, assuming a full range signal. Because the noise at the input is already 40 dB below the signal, only 32 dB are needed additionally. Choosing a cutoff frequency of 70 Hz, the attenuation one decade later (700 Hz) will be 20*n* (dB). Using *n* = 2, the attenuation at 750 Hz will be higher than the minimum required attenuation.

9.2.6 Sample-and-Hold Amplifier

Ideally, sampling a signal requires measuring its instantaneous value at a specific instance of time. In reality, sampling takes some time, different than zero. This time needs to be short enough to ensure that the A/D converter will not detect the change in the value of the signal. The maximal time allowed for sampling decreases when the slope of the signal to measure increases and the number of bits in the A/D converter increases.

Example 9.8

Sampling a 1 kHz sine signal using a 12-bit A/D converter. The change in the signal during the sampling time must not exceed the maximum quantization error. Find the maximum sampling time when the sampling occurs at the peak of the signal and when it crosses zero.

When sampling close to the signal peak value, the condition to meet is

$$\Delta V = A - A \sin\left(\frac{\pi}{2} + 2\pi f t_m\right) < 1\ \text{LSB} = \frac{2A}{2^{12}},$$

from which we can find

$$\frac{\pi}{2} + 2\pi f t_s > 1.5395\ \text{rad}$$

$$t_s < \frac{0.03125\ \text{s}}{2\pi(10^3)} \approx 5\ \mu\text{s}$$

This is a relatively large value of time. However, when sampling at the zero crossing, the derivate of the sine signal $v(t) = A\sin(2\pi f t)$ at that point is $2\pi f A$. If we need for the change in voltage ΔV during a time $\Delta t = t_s$ (sampling time) to be lower than 1 LSB, the condition to meet is

$$\frac{\Delta V}{t_s} = 2\pi f A < \frac{1\ \text{LSB}}{t_s} = \frac{2A}{2^{12}}$$

$$t_s < \frac{1}{\pi 2^{12}(1\ \text{kHz})} = 7.8\ \text{ns}$$

which, as we can see, is a very short time.

If the A/D converter had to digitize (quantify and codify) the input signal in the short amount of time allowed by the sampling process, the resolution obtained would be extremely poor. For this reason, an additional circuit is placed before the A/D converter: the sample-and-hold amplifier (SHA). Figure 9.25 shows the structure of an SHA. It consists of a switch that closes to charge a holding capacitor (C_H) to the value of the input voltage. The amplifier before the switch has high input impedance; the amplifier after the switch has a unity gain and is used to avoid the capacitor discharging during the conversion. *Sampling converters* integrate the SHA.

When the SHA is sampling, it should behave as a unity-gain amplifier. When it is holding, the voltage across the capacitor should be constant.

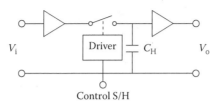

FIGURE 9.25
Functional structure for a sample-and-hold amplifier. The SHA normally integrated in microcontrollers do not have the voltage buffers.

However, the input currents of the output amplifier, the leakage current for the switch, and the leakage from the capacitor itself slowly discharge the capacitor. The droop rate is

$$\frac{dv_c}{dt} = \frac{i_d}{C_H},$$
(9.22)

with i_d being the discharge current (as a result of all the leakage currents in the circuit). The discharge process is slower when C_H increases. During the transition from holding to sampling, the capacitor needs some time to charge to the value of the input voltage. This time is called *acquisition time*. Acquisition time decreases if C_H decreases. There is, therefore, a trade-off between drop rate and acquisition time in choosing the value of C_H. If the conversion time for the A/D converter is short enough, it is possible to use the *track-and-hold mode*. Instead of taking a sample during a very brief period of time and holding its value during a relatively long period of time, in track-and-hold mode the switch is closed for a longer period of time, which allows for the capacitor to charge and follow the changes in the input voltage. At a specific moment, the switch opens and the voltage in the capacitor is digitized during a brief period of time. The time that the switch takes to open from when the order is given to the moment in which the capacitor disconnects itself from the input is called *aperture delay*. This introduces errors in the time at which the sample is really taken because the aperture delay is not constant but subject to erratic and brief changes.

9.2.7 A/D Converters

The A/D converter integrated in microcontrollers, and in most of the peripherals with A/D converters, is based on the successive approximations algorithm shown in Figure 9.26. The input voltage (v_x) is first compared with half of the full-scale voltage ($V_{FS} = V_{ref} = 2^N \times Q$, for an N bit A/D converter). If $v_x > V_{FS}/2$, the most significant bit (MSB) is set to 1 and the compare voltage is increased in $V_{FS}/4$. If $v_x < V_{FS}/2$, the MSB is set to 0

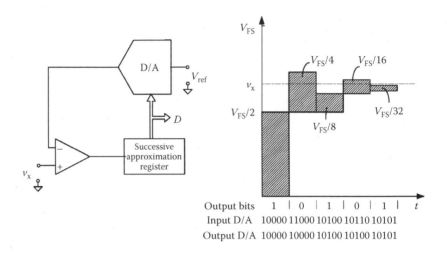

Output bits   1 | 0 | 1 | 0 | 1 | *t*
Input D/A   10000 11000 10100 10110 10101
Output D/A   10000 10000 10100 10100 10101

FIGURE 9.26
Functional structure of a successive approximation D/A converter and decision process. V_{FS}, full-scale voltage.

and the new compare voltage is $V_{FS}/4$. This gives the value of the first bit. To decide the value of the second MSB, the A/D converter proceeds in a similar way: If the result of the second compare is positive ($v_x > V_{compare}$), the bit is set to 1, otherwise it is set to 0. The third compare level will be the previous level plus or minus $V_{FS}/8$ and so on. During this conversion process, v_x must be kept constant. The conversion time will be longer as the number of bits (N) increases. The compare voltages are obtained through a digital-to-analog (D/A) converter (Section 9.6.1).

The relationship between the input voltage for the A/D converter (v_x) and the output code (D) is described by the transfer characteristic shown in Figure 9.3. Some converters are designed with their transfer characteristic shifted to the left in $Q/2$ units, as shown by the discontinuous line in Figure 9.27. This means that the thresholds for the transition between different codes are multiples of $Q/2$ instead of multiples of Q. In reality, the transition between output codes does not always occur for the same input voltage; instead it sometimes occurs for voltages slightly higher, and at other times for voltages slightly lower. The range of voltages that produce the same output code is called *code width*. The *transition threshold* is the voltage that has a 50% probability that the transition may occur for a higher or a lower voltage. In an ideal A/D converter, the code width is 1 LSB for all the codes, and the line that joins the centers of the steps in the transfer characteristic is a straight line with a slope equal to 1 that crosses the origin of the axes.

In a real A/D converter, the line through the center of the steps can have an *offset error* and a *gain error*. The effect of the offset error is to shift

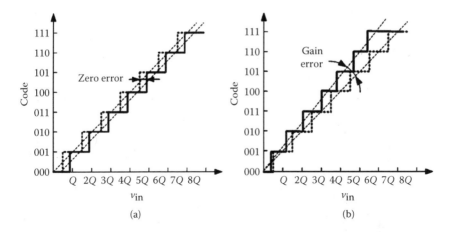

FIGURE 9.27
Transfer characteristic for an A/D converter with (a) zero error and (b) gain error. The ideal transfer characteristic is the bold broken line.

all the transition thresholds in the same direction and by the same value (Figure 9.27a). The effect of gain error, once the offset error has been corrected, is that the slope of the line is different from 1 (Figure 9.27b).

It may also happen that the code width may change between codes. This situation is called *differential nonlinearity* (DNL). DNL is defined as the difference between each code width and the ideal code width of 1 LSB. DNL can be positive or negative. If DNL = −1, it means that either the previous code never happens, or DNL is specified in some extreme conditions in which the previous code would never happen. The global effect of DNL is called *integral nonlinearity* (INL). INL is a measure of the separation between the straight line that crosses the center of the steps compared to the ideal unity slope line once the zero and gain errors have been corrected. If zero and gain errors are corrected by calibration, INL is the factor that limits the accuracy in determining the input voltage that has produced the observed output code. Without using calibration, it is limited by the *absolute error* that is the sum of the zero error, gain error, and nonlinearity error.

For ac signals it is important to know the noise and distortion introduced by the A/D converter. In an ideal A/D converter, the samples would only be affected by the *quantization error*. This is a way of describing the uncertainty in the rms input voltage because all the voltages in the same quantization interval will produce the same output code. Its rms value is $Q/\sqrt{12}$. The rms noise value at the output of a real A/D converter, measured using a standardized process, is always higher. This leads to the definition of the *effective number of bits* (ENOB) as

$$\text{ENOB} = N - \text{lb}\frac{\text{A/D noise}}{Q/\sqrt{12}} = N - \text{lb}\frac{\text{A/D noise}}{\dfrac{V_{FS}}{2^N}/\sqrt{12}} = \text{lb}\frac{V_{FS}/\sqrt{12}}{\text{A/D noise}}. \quad (9.23)$$

In general, for the same value of N, the ENOB is higher for a peripheral A/D than for the A/D integrated in the microcontroller.

The configuration of the input terminals in an A/D converter can be described using the same terms that described the inputs of amplifiers (Section 9.2.2). However, in this case, the configuration is linked to the output codes. If the input is single ended or pseudodifferential, only positive voltages are allowed. In this case, the output code is *unipolar straight binary*. If the input is differential, it accepts positive and negative values referred to the system ground. In this case, the output code is normally a 2-complement binary code because it is easier for numerical calculations. The *full-scale (input) range* (FSR) is equal to V_{ref} for single-ended inputs and $2V_{ref}$ for differential inputs. For a given number of bits (N), the quantization interval for converters with differential input has double the code width than that for single-ended input converters.

According to Equation 9.1, the output of the A/D converter represents the relationship (ratio) between the input voltage (v_x) and the reference voltage (V_{ref}). Therefore, the uncertainty in V_{ref} due to its tolerance, and time and temperature variations will be directly reflected at the output. However, if the input voltage v_x is produced by a sensor powered by the reference voltage such that $v_x = xV_{ref}$ with x proportional to the quantity to measure, and the reference voltage is the same as for the A/D converter, the output of the A/D will not be affected by the uncertainty in V_{ref}. This approach is called *ratio measurement* as opposed to *absolute measurement* when using an independent reference voltage. But it is important to remember that any A/D converter always measures a voltage ratio rather than an absolute voltage.

9.3 The 10-Bit A/D Converter Module in PIC Microcontrollers

9.3.1 Architecture of the Conversion Module

Medium-end PIC microcontrollers use successive approximation A/D converters, normally of 10 bits. A simplified internal structure of these A/D converters is shown in Figure 9.28. The main components of this module are:

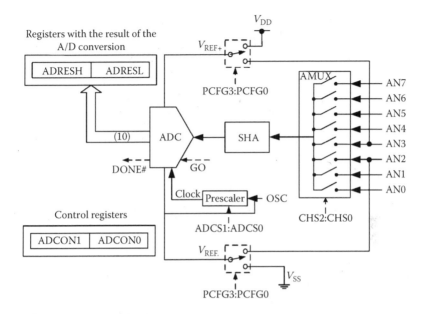

FIGURE 9.28
Functional blocks for the A/D converter in medium-end PIC microcontrollers. A/D, successive approximation analog-to-digital converter; SHA, sample-and-hold amplifier; AMUX, analog multiplexer.

- Analog multiplexer with up to eight input channels
- Sample and hold amplifier without input or output buffers
- 10-bit successive approximation A/D converter
- Registers to control the module (ADCON0 and ADCON1), and registers to store the result of the conversion (ADRESH and ADRESL)

This module can have up to eight analog inputs that are available as alternative functions in the parallel port inputs. The number of analog inputs depends on the specific model of the PIC. For example, the PIC16F873 has five analog inputs that are available in five pins in parallel port A (RA0/AN0, RA1/AN1, RA2/AN2/VREF–, RA3/AN3/VREF+, and RA5/AN4). PICs with more than five analog inputs, such as the PIC16F874 that has eight analog inputs, use three pins from port E for the analog inputs AN5, AN6, and AN7. The selection of channels is done with bits CHS2:CHS0 from the ADCON0 register.

The sample-and-hold amplifier consists of a capacitor (without input or output buffers) that starts to charge when the multiplexer selects the desired channel. The voltage in the capacitor follows the evolution of the input voltage (*track mode*). When the conversion order is given, the capacitor is disconnected from the analog input and the conversion process starts.

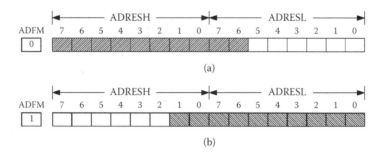

FIGURE 9.29
The result of the A/D conversion can be read in the ADRESH and ADRESL registers with a format specified by bit ADFM in the ADCON1 register. (a) Bit ADFM = 0, resulting in a left alignment. The 8 most significant bits are stored in ADRESH. (b) Bit ADFM = 1, resulting in a right alignment. The 8 least significant bits are stored in ADRESL.

The result of the conversion is stored in registers ADRESH and ADRESL. The set of registers has 16 bits and the result of the conversion is a 10-bit value. Hence, the result of the conversion may be aligned to the right or to the left, as shown in Figure 9.29. Storing the result of the conversion left aligned (Figure 9.29a) is very appropriate to operate the A/D converter as an 8-bit converter, with the result stored in the ADRESH register.

The reference voltage for the A/D converter can be the microcontroller's voltage supply or an external voltage applied between pins AN3/VREF+ and AN2/VREF–. The selection is done with bits PCFG3:PCFG0 in the ADCON1 register. The default selection uses the microcontroller's supply voltage. The A/D conversions are synchronized with a clock signal. This signal can come from the main oscillator clock through a programmable prescaler or by an internal *RC* oscillator working at a fixed frequency. This internal *RC* oscillator is not shown in Figure 9.28. Bits ADCS1 and ADCS0 in the ADCON0 register are used to select the source for the clock and program the prescaler if the clock is taken from the main oscillator. For the A/D converter to continue working while the microcontroller is in sleep mode, it is necessary to select the internal *RC* oscillator.

The conversion is started by activating the control bit GO. When the conversion is finished, the status bit DONE# is activated. In reality, these two flags are implemented using the same bit: GO/DONE# from the ADCON0 register, as shown in Figure 9.30. The programmer must set this bit to 1 to start the conversion. The bit is automatically set to 0 when the conversion is finished and the result is stored in ADRESH and ADRESL. When the conversion is finished, the bit ADIF in the PIR register is also activated to request an interrupt. If bit ADIE in the PIE register is active and the global interrupt system is enabled (bit GIE in the INTON register is 1), the interrupt request becomes effective.

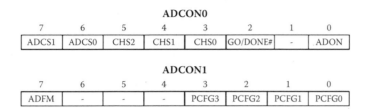

FIGURE 9.30
Registers ADCON0 and ADCON1 in a PIC16F873.

Figure 9.30 shows the bits in the special function registers ADCON0 and ADCON1 in a PIC16F783. These bits are used to control its A/D conversion module. Bits ADCS1:ADCS0 in the ADCON0 register select the clock source for the converter and its frequency, as shown in table 9.1. The analog input is selected with bits CHS2, CHS1, and CHS0. GO/DONE# is the control/status bit to start the conversion and notify when it is finished. With bit ADON = 1, the A/D conversion module in the microcontroller is enabled.

Bit ADFM in the ADCON1 register determines the alignment (right or left) for the conversion result stored in the ADRESH and ADRESL registers. Bits PCFG3:PCFG0 configure the pins of the microcontroller used by the A/D module as analog inputs for the converter or as digital outputs for the corresponding digital ports. Table 9.2 shows the values of these bits for a PIC16F783 microcontroller. After a reset, bits PCFG3:PCFG0 are set to 0 and therefore pins RA5, RA5:RA0 are assigned to the A/D conversion module. To assign them to parallel port A, it is necessary to program the appropriate values in the ADCON1 register.

TABLE 9.1

Selection of Clock Source and Frequency for the A/D Converter Using Bits ADCS1:ADCS0 from the ADCON0 Register

ADCS1:ADCS0	Clock Source	Frequency of the A/D Clock
00	Main oscillator	$F_{OSC}/2$
01	Main oscillator	$F_{OSC}/8$
10	Main oscillator	$F_{OSC}/32$
11	Internal *RC* oscillator	167 kHz to 500 kHz

Note: F_{osc} is the frequency of the main oscillator in the microcontroller. The frequency of the internal *RC* oscillator is fixed with a typical value of 250 kHz, although it can vary between 167 kHz and 500 kHz.

TABLE 9.2

Assignment of Functions to Input Pins in Port A in a PIC16F783 Using the PCFG3:PCFG0 Bits from the ADCON1 Register

PCFG3: PCFG0	AN4 RA5	AN3 RA3	AN2 RA2	AN1 RA1	AN0 RA0	V_{REF+}	V_{REF-}	Number of Analog/ Digital Channels
00x0	A	A	A	A	A	V_{DD}	V_{SS}	5/0
1001	A	A	A	A	A	V_{DD}	V_{SS}	5/0
00x1	A	V_{REF+}	A	A	A	RA3	V_{SS}	4/0
1010	A	V_{REF+}	A	A	A	RA3	V_{SS}	4/0
1x00	A	V_{REF+}	V_{REF-}	A	A	RA3	RA2	3/0
1011	A	V_{REF+}	V_{REF-}	A	A	RA3	RA2	3/0
0100	D	A	D	A	A	V_{DD}	V_{SS}	3/2
0101	D	V_{REF+}	D	A	A	RA3	V_{SS}	2/2
1101	D	V_{REF+}	V_{REF-}	A	A	RA3	RA2	2/1
1110	D	D	D	D	A	V_{DD}	V_{SS}	1/4
1111	D	V_{REF+}	V_{REF-}	D	A	RA3	RA2	1/2
011x	D	D	D	D	D	V_{DD}	V_{SS}	0/5

Note: A, analog input; D, digital input.

9.3.2 A/D Conversion Timing

An analog signal is digitized in two steps: (1) the sample-and-hold process and (2) the A/D conversion process. Each one of these steps needs a specific amount of time. In medium-end PICs, the holding capacitor has a value of 120 pF. The time needed for this capacitor to become charged is called acquisition time (T_{ACQ}). After the capacitor is charged, the 10-bit conversion can start. The time needed for the conversion is called conversion time (T_{CONV}; Figure 9.31).

The manufacturer specifies an acquisition time for medium-end PICs between 10 µs and 20 µs:

$$10 \mu s \leq T_{ADQ} \leq 20 \text{ µs.} \tag{9.24}$$

The wide variation in the acquisition time is due to the lack of the input buffer shown in Figure 9.25. This makes the acquisition time highly dependent on the output (internal) resistance (R_s) of the source signal. The manufacturer recommends that R_s always be lower than 10 kΩ. With $R_s =$ 10 kΩ, T_{ACQ} is 20 µs; whereas with $R_s = 50$ Ω, T_{ACQ} is 10 µs.

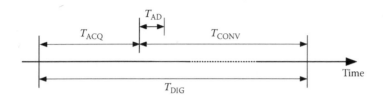

FIGURE 9.31
Times involved in the digitizing process in the A/D module. T_{ACQ}, acquisition time; T_{AD}, time to convert one bit; T_{DIG}, overall data conversion time. T_{ACQ} ranges between 10 µs and 20 µs depending on the value of the output resistance for the source. T_{AD} must be longer than 1.6 µs. The overall data conversion time is 11.5 times longer than T_{AD}.

The manufacturer also specifies that the conversion time for the 10-bit A/D converter in medium-end PICs is

$$T_{CONV} = 11.5 \times T_{AD}, \qquad (9.25)$$

with T_{AD} being the conversion time for a bit.

For the converter to operate correctly, the manufacturer recommends

$$T_{AD} > 1.6 \text{ µs.} \qquad (9.26)$$

With this value, the conversion time for a 10-bit A/D converter is equal to 18.4 µs.

The value of T_{AD} is equal to the period of the clock in the A/D converter. Because the clock for the A/D converter can be taken from the main oscillator, the frequency of this oscillator must be chosen so that it meets the condition from Equation 9.26. Table 9.3 shows the maximum frequency value for the main oscillator in the microcontroller for the different possible configurations.

The digitizing time for the analog signal (T_{DIG}) is the sum of the acquisition and conversion time:

$$T_{DIG} = T_{ACQ} + T_{CONV}. \qquad (9.27)$$

Based on the previous data, it can be seen that for these PIC microcontrollers the lowest value for T_{DIG} ranges between 20.4 µs and 38.4 µs.

When an analog signal is digitized periodically with a sampling period T_s, the sampling frequency is $F_s = 1/T_s$. In this case, T_s must be equal to or higher than the time needed for the digitizing process. The manufacturer recommends waiting $2T_{AD}$ seconds before starting a new conversion. Therefore,

$$T_s \geq (T_{DIG} + 2\,T_{AD}). \qquad (9.28)$$

This expression can be rearranged to find the limit of the sampling frequency as

$$F_S \leq \frac{1}{T_{DIG} + 2T_{AD}}.$$

(9.29)

Example 9.9

Digitizing time for a 10-bit A/D conversion in a PIC16F783 with a main oscillator at 4 MHz.

With $F_{osc} = 4$ MHz, and considering Table 9.3, the configuration that is chosen is ADCS1:ADCS0 = 01, resulting in $T_{AD} = 8/F_{osc} = 2.0$ µs. This value meets the requirements from Equation 9.26. The 10-bit A/D conversion time is therefore

$$T_{CONV} = 11.5 \times T_{AD} = 11.5 \times 2.0 \text{ µs} = 23 \text{ µs}.$$

Considering the worst case for the acquisition time ($T_{ACQ} = 20$ µs) that corresponds to a signal with a source resistance of 10 kΩ, the resulting sampling time is

$$T_{DIG} = T_{ACQ} + T_{CONV} = 20 \text{ µs} + 23 \text{ µs} = 43 \text{ µs}.$$

If the A/D conversion is carried out periodically, the sampling frequency must be equal to or less than $1/(T_{DIG} + T_{AD})$, resulting in $F_s \leq 22.222$ kHz.

Considering the best case for the acquisition time ($T_{ACQ} = 10$ µs) that corresponds to a signal with a source resistance of 50 Ω, the resulting sampling time is

$$T_{DIG} = T_{ACQ} + T_{CONV} = 10 \text{ µs} + 23 \text{ µs} = 33 \text{ µs}.$$

This results in a maximum sampling frequency of 28.571 kHz.

TABLE 9.3

Maximal FOSC in the Microcontroller in Order to Have TAD =1.6 µs for the Different Possible Configurations.

ADCS1:ADCS0 in ADCON0	T_{AD}	F_{osc}	F_{osc} (for $T_{AD} = 1.6$ µs)
00	$2/F_{osc}$	$2/T_{AD}$	1.25 MHz
01	$8/F_{osc}$	$8/T_{AD}$	5 MHz
10	$32/F_{osc}$	$32/T_{AD}$	20 MHz
11	2 µs to 6 µs	—	—

9.3.3 A/D Conversion Module Programming

The A/D conversion module can be serviced by using the polling or interrupt techniques. The steps necessary to measure analog voltages in an input channel are described in the following.

1. Configure the A/D conversion module.
   - Configure the pins in ports A and C as analog inputs, reference voltage, or digital I/O storing the appropriate values in bits PCFG3:PCFG0 in the ADCON1, TRISA, and TRISC registers.
   - Configure the format for the conversion result with bit ADFB in register ADCON1.
   - Select the source of the clock used by the conversion module and the bit conversion time (T_{AD}) using bits ADCS1:ADCS0 in the ADCON0 register.
   - Select the analog input channel using bits CHS2:CHS0 from ADCON0.
   - Activate the A/D module with bit ADON from the ADCON0 register.
2. If the A/D module is serviced by interrupts, configure the interrupt in the A/D module.
   - Set bit ADIF in the PIR register to 0. This is the interrupt flag.
   - Enable the A/D converter interrupt by setting bit ADIE in the PIE register to 1.
   - Enable the general interrupt system in the PIC by setting bit GIE in the INTCON register to 1.
3. Wait the required acquisition time (T_{ACQ}).
4. Start the A/D conversion by setting to 1 bit GO/DONE# in the ADCON0 register.
5. Wait for the A/D conversion to be complete:
   - If using polling service: Wait for the bit GO/DONE# in the ADCON0 register to be 0 or the bit ADIF in the PIR register to be 1.
   - If using interrupt service: Wait for the A/D converter interrupt.
6. Read the result of the conversion in the ADRESH and ADRESL registers. Set bit ADIF to 0 if needed.
7. To acquire another sample, repeat steps 1 or 2 as needed. Wait for at least $2T_{AD}$ before acquiring a new sample.

Example 9.10 shows how to program the A/D module to acquire analog signals using polling input.

Example 9.10

Programming the A/D conversion module in a PIC16F873 with its main oscillator at 4 MHz to acquire the signal from any of its five analog input channels.
The process consists of three parts:

1. Initialization module: Label Init that configures port A inputs as analog inputs and establishes the format for the result of the A/D conversion.
2. Subroutine channel: Configures the word to store in the ADCON0 register depending on the channel to measure.
3. Subroutine measure: Receives in register W the number of the channel to measure and return, also in W, the result of the measure. This subroutine gives the number of the channel to measure to the subroutine channel and receives the appropriate word to store in the ADCON0 register. After the configuration, it generates a delay for the recommended acquisition time. Once this time has elapsed, it starts the A/D conversion by setting the bit GO/DONE# in ADCON0 to 1. The subroutine waits until this bit is set to 0, thus indicating that the A/D conversion has finished. Finally it returns the 8 MSBs of the result in register W.

```
; Programming the 10 bit A/D converter.
; Fosc = 4 MHz
      List        p = 16F873
      include   «P16F873.INC»
AUX   equ         0x20              ; Auxiliary variable.
      org         0x00
      goto        Init
      org         0x04
      retfie
Init:
      BSF         STATUS, RP0   ; Select bank 1.
      movlw       0xff          ; W with ffh.
      movwf       TRISA         ; PORTA as input.
      clrf        ADCON1        ; All inputs in PORTA are
                                ; analog and
                                ; result of conversion aligned
                                ; left.
;
; Write here main program.
;
; Subroutine Measure:
; This subroutine carries out the 10-bit A/D conversion for
; the analog channel.
; The subroutine selects the desired channel and waits until
; the result is ready
; in registers ADRESH and ADRESL. It returns in W the 8 most
; significant bits
; of the conversion (value of ADRESH).
; This subroutine assumes a bit conversion time Tad= 2
; microseconds and
; a main clock of 4 MHz for the PIC
; Inputs: in W the number of the channel to measure.
; Outputs: in W the 8 bit result of the  measurement.
;
Measure:
      nop                             ; Wait 2Tad = 4 microseconds
      nop
```

```
        nop
        nop
        bcf      STATUS, RP0      ; Select bank 0.
        call     Channel          ; Select the word to store in
        movwf    ADCON0           ; ADCON0 depending on the
                                  ; channel.
        call     Del10us          ; Wait for an acquisition time
                                  ; of 10 microseconds.
        bsf      ADCON0, GO       ; Start the A/D conversion.
Measure01:
        btfsc    ADCON0, GO       ; Conversion finished?
        goto     Measure01        ; No - continue waiting.
        movf     ADRESH, W        ; Yes - Store results in W.
        bcf      ADCON0, ADON     ; Disable the A/D converter.
        return                    ; Return.
; Subroutine Channel.
; This subroutine receives in W the number of the channel and
; returns in W
; the word to store in ADCON0 to enable the A/D converter,
; select the input channel and choose the clock forthe A/D
; conversion module.
; The clock in the A/D conversion module has been selected at
; Fosc/8.
; With Fosc = 4 MHz, the bit conversion time is Tad = 2
; microseconds.
;
Channel:
        addwf    PCL, f
        retlw    41h              ; Word to select channel 0.
        retlw    49h              ; Word to select channel 1.
        retlw    51h              ; Word to select channel 2.
        retlw    59h              ; Word to select channel 3.
        retlw 61h                 ; Word to select channel 4.
; Subroutine Del10us.
; This subroutine delays for more than 10 microseconds.
;
Del10us:
        movlw    .3
        movwf    AUX
Del01:
        decfsz   AUX
        goto     Del01
        return
        end
```

9.4 Calibration

To correctly interpret the code obtained in the A/D conversion in terms of the input voltage, it is necessary to know the real transfer characteristic between the front-end and the A/D converter. Ideally, if the quantization is excluded, this function is a straight line with a slope equal to the gain of the system:

$$D = G \times v_x + V_0. \qquad (9.30)$$

However, the real transfer characteristic in some given conditions (voltage supply, temperature, frequency) can have a slope different from G and an offset different from V_0 (Figure 9.32a). The real values of the gain and offset can be found by applying two known input voltages V_1 and V_2 and relating them to the readings D_1 and D_2 that they originate. With this, it is possible to find the relationship between any code D given by the A/D converter and the voltage v_x that produced it as:

$$v_x = \frac{V_2 - V_1}{D_2 - D_1}(D - D_1) + V_1. \tag{9.31}$$

This process is called *calibration* and allows correcting for constant deviations between the ideal and real responses. The voltage V_1 is normally chosen as 0 V and the voltage V_2 as the voltage that produces the full-scale voltage at the input of the converter, V_{FS}. With $V_0 = 0$ V, then $V_2 = V_{FS}/G = V_{ref}/G$. Figure 9.32b shows how to implement this calibration process when the input currents for the multiplexers used (or the output resistance for the channels) are low enough so that their influence will be below 1 LSB. Otherwise, it is necessary to include a resistor equal to the value of the resistance of the output channels in the connection to ground, and also in the connection from the reference voltage. It is important to note that this only calibrates the part of the system between the application of the known voltages and their output. If the transfer characteristic is a straight line with a slope very close to the ideal slope, it is possible to calibrate with a single value, normally with $V_1 = 0$ V.

If the actual response of the system is not a straight line, the input range can be divided into two or more contiguous subranges and the calibration process applied to each of them. Each subrange should be a straight line, thus allowing it to be calibrated using the previous process. Given a reading D, the first step is to determine what voltage range originated

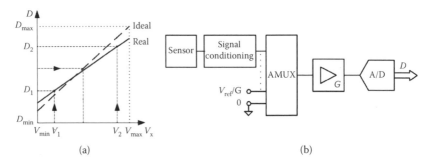

(a)   (b)

FIGURE 9.32
Calibration of a data acquisition system using two known voltages (V_1 and V_2) to determine the real transfer characteristic.

that specific reading. The potential nonlinearity can be determined by applying a known voltage V_3 so $V_1 < V_3 < V_2$. If the value obtained using Equation 9.31 is not V_3, the response is nonlinear.

9.5 Direct Sensor–Microcontroller Interface

Analog signals that carry information in their time instead of in their amplitude can be digitized using a simple timer or digital counter like the ones available in many microcontrollers, as long as their inputs are Schmitt trigger (ST) inputs. Otherwise, it is necessary to incorporate external ST devices. The duration of the signal is determined by counting cycles from the internal clock between the edges of the signal. If these edges are not steep and there is noise with an amplitude higher than the hysteresis cycle for the ST, the beginning and end of the counting process can be erroneous.

The circuit shown in Figure 9.33 can be used to encode information in the duration of a signal when using a resistive sensor with a value R_x. The capacitor is first charged to the high output value (V_{OH}) in pin 1, while pin P is kept in a high impedance state. Afterward, pin 1 is kept at a high impedance level, while pin P is set to the low-output voltage (V_{OL}). This makes the capacitor discharge through R_x. The discharge time until the low threshold is detected by the Schmitt trigger circuit (V_{TL}) and is proportional to the product $R_x C$ as well as to the voltages V_{OH} and V_{TL}. The same approach could be used to measure the charge time instead of the discharge time. However, the low threshold for the Schmitt trigger circuit is less affected by noise than the high threshold.

The calibration circuit shown in Figure 9.34a can be used to obtain a reading for R_x independent of the values of C, V_{OH}, and V_{TL}, with R_{c1} and

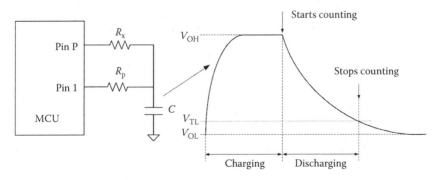

FIGURE 9.33
Measuring an unknown resistance based on the time needed to discharge a capacitor. R_p limits the capacitor charging current to less than 20 mA.

R_{c2} being known resistors. The measurement algorithm is the same as described for Figure 9.33, although in this case the cycle is repeated three times, each time discharging the capacitor through a different resistor. The unknown resistance R_x can be found as

$$R_x = \frac{N_x - N_{c1}}{N_{c2} - N_{c1}} \left(R_{c2} - R_{c1} \right) + R_{c1} ,$$ (9.32)

with N_x, N_{c1}, and N_{c2} being the discharge times, measured in clock cycles, through the resistors R_x, R_{c1}, and R_{c2}. Although one of the known resistors could be a short circuit, it is better to use resistors with values close to the extreme possible values for R_x.

Applying the same method to the bridge of sensors shown in figure 9.34b results in

$$\frac{N_1 - N_3}{N_2} = \frac{\Delta R}{R_0} ,$$ (9.33)

with N_1, N_2, and N_3 being the readings obtained when discharging the capacitor through pins 2, 3, and 4, respectively.

The variable resistance and the sensor bridge shown in Figure 9.34 are connected to the microcontroller without any external element other than the capacitor. This approach is called *direct interfacing* because there are no integrated circuits between the sensor and the microcontroller.

Microcontrollers with an internal A/D converter can be directly connected to sensors in circuits whose output is a voltage, provided that the resolution of the converter is adequate to the voltage range. However, if the sensor is a resistor bridge, the A/D converter must have a differential input. For the circuits shown in Figure 9.34, an A/D converter is not necessary as the measurement of the unknown resistance is done with a timer. The method described for Figure 9.33 can also be used to measure capacitive sensors, switching the connections of the capacitor and resistor. The calibration circuit must then include two known capacitors. One of them can be an open circuit ($C = 0$), as shown in Figure 9.35. With this, the parasitic capacitances between node N and the ground are charged in the first phase and discharged through R. Choosing an R higher than 1 MΩ to have longer discharge times makes the circuit extremely sensitive to interference even when the connections to C_x are very short.

This method for measuring resistances and capacitances has two inherent limitations: The first one comes from the quantization in the counting process, as the result can only be an integer multiple of the clock cycle. The second limitation arises from the uncertainty in crossing the detection

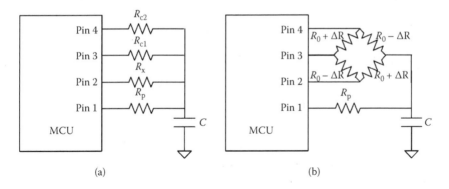

(a) (b)

FIGURE 9.34
(a) If two known resistances are measured with the method shown in Figure 9.33, it is possible to determine R_x without knowing the value of C or the output voltage values in the microcontroller. (b) A sensor bridge can be connected as a circuit with three input terminals and one output terminal and use the same measurement method as in (a).

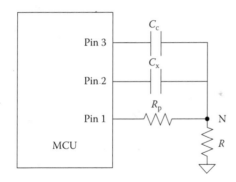

FIGURE 9.35
Measuring an unknown capacitor based on its discharge time through resistor R. R_p limits the charging current for the capacitor to less than 20 mA. If two known capacitances (open circuit and C_c) are measured with the same method, we can determine C_x without knowing R or the output voltages for the microcontroller.

threshold due to noise superimposed to the discharge voltage or to the threshold voltage V_{TL}.

The limitation from the quantization process depends on the method used to detect when the discharging signal crosses the threshold voltage for the Schmitt trigger (Section 5.1.2). If the detection is done by polling, the event may occur immediately after the microcontroller is polled. This results in an uncertainty that can vary between one clock cycle and the number of clock cycles between polls. If using interrupts, the uncertainty can vary between one clock cycle and the number of cycles for the longest instruction that needs to be finished before servicing the interrupt. If the microcontroller has a capture module (Section 6.2), the value of the timer connected to the discharge time is captured when the threshold V_{TL} is detected.

The effect of the noise in triggering the Schmitt trigger circuit depends on the slope of the signal at that point. Assuming constant amplitude for the noise, the error in detecting the trigger time will increase when the slope of the signal decreases. For this reason, using a large RC time constant to increase the number of counts will result in an increased error due to the noise as the signal will be slower. However, if the RC constant is selected small in order to have a signal with a faster slope, then the number of counts will be smaller and the quantization will have an increased effect on the error. The optimal value for thesconstant depends on the noise levels for the threshold voltage and the discharging signal. For a printed circuit board designed correctly, if power supply voltage is decoupled, RC can be between 1 ms and 3 ms, resulting in a resolution between 10 bits and 12 bits. If the noise effect rather than quantization predominates, the resolution can be increased by averaging several readings.

9.6 Analog Back-End

To obtain an analog signal at the output of a microcontroller, it is necessary to reverse the process described for the front-end: one or several signals must be reconstructed from their values at given times. These functions are carried out by the analog back-end.

9.6.1 D/A Converters

A digital-to-analog (D/A) converter (DAC), shown in Figure 9.36a, outputs a voltage or a current whose amplitude corresponds to the digital code at the input $(B_{n-1}, B_{n-2}, ..., B_1, B_0)$. The output voltage or current is a fraction of the reference voltage:

$$V_o = V_{ref} \left(\frac{B_{n-1}}{2} + \frac{B_{n-2}}{2^2} + ... + \frac{B_1}{2^{n-1}} + \frac{B_0}{2^n} \right). \qquad (9.34)$$

Each bit B_i can be 0 or 1. Therefore, this function can be seen as the product of an analog voltage V_{ref} and a digital code. Figure 9.36b shows a standard circuit to implement this function. Each bit opens or closes a switch whose position in the R–$2R$ network is farther away from the output as the significance of the bit increases. The D/A converters integrated in microcontrollers use a C–$2C$ capacitive network instead of a resistive network, as capacitors are easier to integrate than resistors. The connection of the

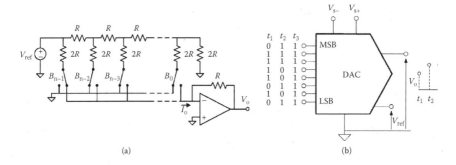

FIGURE 9.36
D/A converter: function and standard circuits based on an *R*–2*R* network.

C–2C is slightly different than the connection for the *R–2R* network, but the final effect is the same.

The ideal and real transfer characteristics of a D/A converter can be described using the same parameters that describe the transfer characteristics for an A/D converter (Figure 9.27).

9.6.2 Analog Demultiplexing

Some D/A converters have multiple channels, up to 16 channels with 16 bits to produce several analog signals simultaneously. However, if the frequency of these signals is not too high, it is possible to use a fast single D/A converter and an analog demultiplexer, as shown in Figure 9.37. This analog demultiplexer carries out the inverse function of the input multiplexers shown in Figures 9.9 and 9.10.

The output voltage for each channel is connected to a sample-and-hold amplifier by closing the appropriate switch in the demultiplexer. The holding capacitor for each channel must be large enough so it will not discharge significantly before its voltage is refreshed or changed from the D/A converter. However, larger capacitance values will need longer charging times. Analog demultiplexing can be done with the same multiplexers described in Section 9.2.4 because, unlike digital multiplexers, analog multiplexers are reversible (Figure 9.21).

9.6.3 Extrapolation Methods

The voltage at the output of the D/A converter corresponds to its input code. Therefore, the output voltage will be kept constant until the D/A updates the value of its input. If the D/A converter is connected to a sample-and-hold amplifier, the output of this amplifier will be constant until it samples another value. This is called a *zero-order hold* (ZOH) extrapolation because the output is constant until a different value comes. Because the

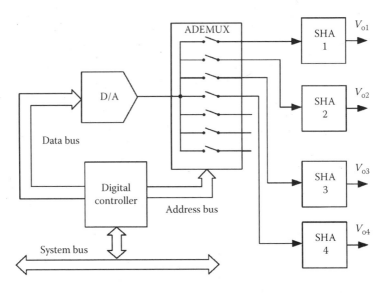

FIGURE 9.37
Analog demultiplexing to obtain several analog signals with a single D/A converter.

output signal looks like a staircase, to have a smooth signal it is necessary to update the output at a higher rate as shown in Figure 9.38. The modulus of the transfer function for the zero-order extrapolation is

$$\left|H(f)\right| = \left|T\frac{\sin \pi fT}{\pi fT}\right|, \tag{9.35}$$

with T being the holding time. For a sine signal with an amplitude of A volts and frequency f, in order for the difference between the ideal and the staircase signal to be less than, say, $A/2^{10}$, it is necessary that $fT < 41$. This means that the sine signal should be approximated with at least 41 steps.

9.6.4 PWM Outputs

The dc value of a pulse width modulated signal, such as the one depicted in Figure 9.39, depends on its duty cycle (Section 6.2.3) as

$$V_0 = \left(V_{OH} - V_{OL}\right)\frac{T_{ON}}{T} \approx V_{DD}\frac{T_{ON}}{T}, \tag{9.36}$$

with T being the pulse period. This equation assumes that the pulse high voltage is V_{DD} and its low voltage is 0 V. In practice, depending on the type of microcontroller used and its output currents, the high voltage can be

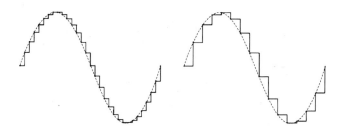

FIGURE 9.38
Effect of the frequency of the samples on the waveform for the reconstructed signal when using a zero-order hold. For better similarity between the reconstructed signal and the ideal signal, it is necessary to have a higher number of samples (left). Few samples per unit of time yield a signal with poor resemblance to the ideal signal (right).

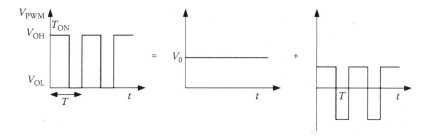

FIGURE 9.39
A PWM can be produced as the sum of a dc signal and a square signal without dc value, which can be seen as the sum of several harmonic sine signals. The amplitude of the first sine signal will be $(V_{OH} - V_{OL})2/\pi$.

as low as $V_{DD} - 0.7$ V and the low voltage can be as high as 0.6 V. In any case, by adjusting the duty cycle, it is possible to obtain any desired dc voltage. The stability of this dc voltage directly depends on the stability of V_{DD}. If the current sourced by the output gates is not high enough, we can connect a voltage comparator between the PWM output and the low-pass filter. To achieve an output with a higher voltage, the comparator should be open-collector or open-drain.

If the PWM signal is filtered using a first-order low-pass filter with a cutoff frequency f_c, the peak-to-peak voltage for the first harmonic at the output of the filter is

$$V_1 = \frac{2V_{DD}}{\pi} \frac{1}{\sqrt{1 + \left(\dfrac{f}{f_c}\right)^2}}.$$ (9.37)

Higher-order harmonics have higher attenuation, decreasing with the square of the order of the filter. If the ripple due to the first harmonic has to be less than 1 LSB/2^q for a system with N bits, the cutoff frequency of the filter must meet

$$f_c < \sqrt{\frac{\pi}{2}} \frac{f}{2^{N+q} - 1} \approx \frac{1,25 f}{2^{N+q}}. \tag{9.38}$$

For example, if $f = 20$ kHz and $N = 8$, the cutoff frequency has to be less than 25 Hz to achieve a ripple lower than one-fourth LSB. Active filters have lower output impedances compared to passive filters, but they add an offset voltage to the output voltage. They are also limited in the range of values for the output voltage.

In addition to the effects of the ripple, the resolution in the dc output voltage is also limited due to the resolution for T_{ON}. According to Equation 9.36, the output voltage can be lowered by increasing T, but this implies reducing the cutoff frequency for the low-pass filter. This in turn causes a slower transient response when changing dc output values.

When the desired output is a sine signal with a dc offset, this can be implemented by the modulating signal for the PWM signal having the desired output frequency. In this case, however, the output filter must be an active filter, order 3 or higher, to have a ripple compatible with an 8-bit system or higher. If the maximal frequency for the PWM signal is about 20 kHz, this method can generate signals up to 1 kHz. These signals can be used as test signals or for acoustical communication.

9.6.5 Output Protections

The voltage, current, and power levels that microcontrollers can handle are limited to relatively low values in the range of $V_{SS} - 0.3$ V to $V_{DD} + 0.3$ V and ± 25 mA. Therefore, it is necessary to use external drivers when trying to operate on loads that require higher power levels. It is also necessary to protect the microcontroller against higher values of voltages and current. Section 9.2.3 describes the protections at the input to protect the microcontroller against anomalous situations. However, even normal operating conditions can result in dangerous voltage and current levels in the output pins. When the current through an inductive load is suddenly interrupted, the inertia of the current to continue moving creates a voltage difference across the open circuit of the switch. This voltage can be around 20 V to 30 V even for small currents and inductances. This overvoltage can be prevented by placing a diode or a varistor in parallel with the load, as shown in Figure 9.40a.

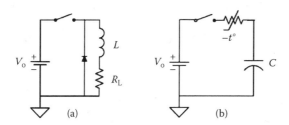

FIGURE 9.40

Output protections for (a) inductive load: voltage limiter, and (b) capacitive load: current limiter. The switch represents the action from 0 to 1 (closing) and 1 to 0 (opening) in the output gate.

When the switch opens (for example, moves from 1 to 0) the current that was flowing through the inductance will flow through the diode until it extinguishes.

The problem with capacitive loads arises when they are turned on, because the initial current through the capacitor is only limited by the internal resistance of the load. The solution in this case is to place a resistor with a negative temperature coefficient (NTC thermistor) in series with the load, as shown in Figure 9.40b. The initial value of this resistance is high, at least 250 Ω, which is enough to limit the initial current. As the current flows, the NTC thermistor heats up and its value diminishes; the capacitor can charge quickly.

Appendix: Acronyms

A/D	analog-to-digital (converter)
ACC	accumulator
ADEMUX	analog demultiplexer
AFE	analog front-end
ALU	arithmetic and logic unit
AMUX	analog multiplexer
ASCII	American Standard Code for Information Interchange
BISYNC	Binary Synchronous Communication
BOR	brown-out reset
CCP	compare/capture/PWM
CISC	complex instruction set computer
CMOS	complementary metal-oxide semiconductor
CMRR	common mode rejection ratio
CPU	central processing unit
CSMA/CD	Carrier Sense Multiple Access with Collision Detection
D/A	digital-to-analog (converter)
DCE	data communication equipment
DIP	dual in-line package
DMA	direct memory access
DNL	differential nonlinearity
DR	dynamic range
DSP	digital signal processor
DTE	data terminal equipment
EEPROM	electrical erasable programmable read-only memory
EIA	Electronic Industries Alliance
ENOB	effective number of bits
EPROM	erasable programmable read-only memory
FIFO	first in, first out
FPGA	field programmable gate array
FS	full scale
FSR	full-scale (input) range
GIE	global interrupt enable
GPR	general purpose register
HDLC	High-Level Data Link Control
I²C	inter-integrated circuit
ICSP	In-Circuit Serial Programming
IEC	International Electrotechnical Commission
INL	integral non-linearity
IR	instruction register
LCD	liquid-crystal display

LED	light-emitting diode
LIFO	last in, first out
LPF	low-pass filter
LSB	least significant bit
MC	machine cycle
MODEM	modulator–demodulator
MOV	metal-oxide varistor (variable resistor)
MPASM	macro assembler for PIC microcontrollers
MSB	most significant bit
MSSP	master synchronous serial port
NTC	negative temperature coefficient
OST	oscillator start-up timer
OTP	one-time programmable
PC	program counter
PCON	power control
PIC	programmable integrated circuit
PLD	programmable logic devices
POR	power-on reset
PSP	parallel slave port
PTC	positive temperature coefficient
PWM	pulse width modulation
PWRT	power-up timer
RAM	random-access memory
RISC	reduced instruction set computer
ROM	read-only memory
RS-232C	Recommended Standard 232, Revision C
RTC	real-time clock
SCI	serial communication interface
SCL	serial clock line
SDA	serial data line
SDLC	Synchronous Data Link Control
SFR	special function register
SI	International System of Units
SP	stack pointer
SPI	serial peripheral interface
SPP	slave parallel port
SSP	synchronous serial port
ST	Schmitt trigger
STATUS	status register or bit
USART	universal synchronous asynchronous transmitter receiver
USB	universal serial bus
W	working register
WDT	watchdog timer
XTAL	crystal
ZOH	zero-order hold

Bibliography

Baker, B. *A Baker's Dozen: Real Analog Solutions for Digital Designers*. Burlington, MA: Newnes, 2005.

Bowling, S. *Understanding A/D Converter Performance Specifications*, AN693. Chandler, AZ: Microchip Technology, Inc., 2002.

Cravotta, R. *The 32nd Annual Microprocessor Directory*. EDN, August 4, 2005.

EDN's 2005 Microprocessor/Microcontroller Directory. EDN, August 5, 2004.

Embedded Control Handbook Update 2000. Chandler, AZ: Microchip Technology, Inc., 1999.

Fundamentals of RS–232 Serial Communications, Application Note 83. Dallas Semiconductor. March 29, 2001.

The I²C-Bus Specification, Version 2.1. Philips Semiconductors, January 2000.

Irazabal, J.-M., and S. Blozis. *I²C Manual*, AN10216-01. Philips Semiconductors, March 24, 2003.

MCS-51™ Microcontroller Family User's Manual. Mt. Prospect, IL: Intel Corporation, February 1994.

MPLAB® IDE User's Guide. Chandler, AZ: Microchip Technology, Inc., 2005.

Pallas-Areny, R., and J. G. Webster, *Analog Signal Processing*. New York: John Wiley & Sons, 1999.

Pallas-Areny, R., and J. G. Webster, *Sensors and Signal Conditioning*, 2nd ed. New York: John Wiley & Sons, 2001.

Pardo Carpio, F., *Edu-PIC User's Manual*. Universitat de València, http://tapec.uv.es/edupic, July 2002.

Peatman, J. B. *Design with Microcontrollers*. New York: McGraw Hill, 1988.

Peatman, J. B. *Design with PIC Microcontrollers*. Upper Saddle River, NJ: Prentice-Hall, 1997.

PIC16F87X Data Sheet 28/40-Pin 8-Bit CMOS FLASH Microcontrollers. Chandler, AZ: Microchip Technology, Inc., 2001.

PICmicro™ Mid-Range MCU Family Reference Manual. Chandler, AZ: Microchip Technology, Inc., December 1997.

Reverter, F., and R. Pallàs-Areny, *Direct Sensor-to-Microcontroller Interface Circuits: Design and Characterization*. Barcelona, Spain: Marcombo, 2005.

Richey, R. *How to Implement ICSP™ Using PIC16CXXX OTP MCUs*. Chandler, AZ: Microchip Technology, Inc., 1999.

Wharton, J. *An Introduction to the Intel® MCS-51™ Single-Chip Microcomputer Family*, AP-69. Intel Corporation, 1980.

Index

A